Research Assessment Framework for Global Universities 2020

Research Assessment Framework for Global Universities 2020

Kiat Seng Yeo

Zhiren Yang

Andy Chan

World Scientific

NEW JERSEY · LONDON · SINGAPORE · BEIJING · SHANGHAI · HONG KONG · TAIPEI · CHENNAI · TOKYO

Published by

World Scientific Publishing Co. Pte. Ltd.

5 Toh Tuck Link, Singapore 596224

USA office: 27 Warren Street, Suite 401-402, Hackensack, NJ 07601

UK office: 57 Shelton Street, Covent Garden, London WC2H 9HE

British Library Cataloguing-in-Publication Data
A catalogue record for this book is available from the British Library.

RESEARCH ASSESSMENT FRAMEWORK FOR GLOBAL UNIVERSITIES 2020

ISBN 978-981-122-952-7 (paperback)
ISBN 978-981-122-957-2 (ebook for institutions)
ISBN 978-981-122-958-9 (ebook for individuals)

For any available supplementary material, please visit
https://www.worldscientific.com/worldscibooks/10.1142/12078#t=suppl

World University Research Rankings 2020

www.worldresearchranking.com

Executive Summary

The World University Research Rankings 2020 is a novel methodology designed to evaluate research multi-disciplinarity, complemented by indicators to gauge research impact and research collaborative-ness. Together, these three key aspects make up the composite index for the world university research rankings methodology.

The aim of this evidence-based new index is to inform universities objectively on their world research standing; provide insights into their relative strengths and weaknesses; raise awareness and promote collaborative and multi-disciplinary research. The indicators that were chosen reflect this aspirational goal to help universities better understand how they can seize opportunities in this increasingly complex and inter-connected world, and relook at how universities' research excellence can be assessed holistically with respect to how they are able to harness the synergies created from bringing together multi-disciplinary and collaborative teams to sustain and elevate research impact.

This new index is designed through a novel and fresh methodology from the existing ones (i.e. QS, THE, ARWU) to evaluate the three key components – research multi-disciplinarity, research impact, and research collaborative-ness. Together, these form a new and differentiated composite index under this research assessment framework for universities. Data is derived objectively based on OECD's schema in Web of Science's InCites through a 10-year period from 2009 to 2018.

Contents

1. Introduction

As the global environment and societal challenges grow increasingly complex and multi-dimensional, such as those pertaining to climate change, ageing population and healthcare associated issues, sustainability and urbanization, universities are impelled to adopt a more multi-disciplinary and collaborative approach towards research in pursuit of more holistic and efficacious solutions.

Concurrently, in the pursuit of academic and research excellence, universities have increasingly adopted performance benchmarking with the aid of various tools, including but not limited to global university rankings, among which the most prominent are those by Quacquarelli Symonds (QS), Times Higher Education (THE) and Academic Ranking of World Universities (ARWU).

QS evaluates universities based on the following six indicators[1]: Academic Reputation - 40%, Employer Reputation - 10%, Faculty/Student Ratio - 20%, Citations per faculty - 20%, International Faculty Ratio - 5%, International Student Ratio - 5%.

THE evaluates universities based on the following thirteen indicators[2]: Teaching reputation - 15%, Staff to Student ratio - 4.5%, Doctorate to Bachelor ratio - 2.25%, Doctorate awarded to academic staff ratio - 6%, Institutional income - 2.25%, Research reputation - 18%, Research productivity - 6%, Research income - 6%, Citations - 30%, Proportion of international students - 2.5%, Proportion of international staff - 2.5%, International collaboration - 2.5%, Industry Income - 2.5%.

ARWU evaluates universities based on the following six indicators[3]: Alumni of an institution winning Nobel Prizes and Fields Medals - 10%, Staff of an institution winning Nobel Prizes and Fields Medals - 20%, Highly Cited Researchers - 20%, Papers published in Nature and Science - 20%, Papers indexed in Science Citation Index-Expanded and Social Science Citation Index - 20%, Per capita academic performance of an institution - 10%.

The five most notable observations of these university rankings are:

(1) Surveys of opinions play a significant role;
(2) The top of these rankings is dominated by older, bigger, more renowned universities from a few countries;
(3) Methodologies may not be easily replicated to quickly verify results;
(4) Negligible weightage for industry and international collaboration; and
(5) Absence of indicator measuring research multi-disciplinarity.

Bahram Bekhradnia[4] opined that "the international surveys of reputation should be dropped - methodologically they are flawed, effectively they only measure research performance and they skew the results in favour of a small number of institutions…Ranking providers should acknowledge that their rankings effectively only measure research performance; and the sector more widely should acknowledge and publicize this fact".

Philip G. Altbach[5] noted that "most experts are highly critical of the reliability of simply asking a rather unrandom group of educators and others involved with the academic enterprise for their opinions…Some of AWRU's criteria clearly privilege older, prestigious Western universities - particularly those that have produced or can attract Nobel prizewinners".

Bowman et al.[6] also noted significant issues and biases with reputational surveys including but not limited to "strong evidence for an anchoring effect on assessments of institutional reputation".

Mmantsetsa Marope et al.[7] opined that "On the positive side, rankings address the growing demand for accessible, manageably packaged and relatively simple information on the 'quality of higher education institutions'. This demand is greatly fueled by the need to make informed choices of universities, within a context of massification of higher education and the widely growing diversity of providers. The growing base of stakeholders also fuels this demand. Students use the information to choose where to study, just as their parents and governments use it to place children in the 'best' universities. Donors use rankings to best place their endowments so as to realize the best potential value for their investments. The private sector likewise uses them to identify promising partner institutions in higher education, as do faculty when identifying research collaborators. Policy-makers and universities themselves turn to rankings to learn more about the strengths of their higher education institutions and to identify potential areas for improvement. Governments also often use them to gauge the global standing of their institutions and therefore their competitiveness."

K.S. Yeo et al.[8] also viewed that "The ranking of worldwide universities to assess their research performance, especially in a specific field, allows each university to be aware of its standing in the global arena. Armed with this knowledge, the university's management can then set a visible and yet reasonable goal for the university to strive forward diligently as well as devising strategic plans to outperform its competitors in the near future."

As indicated in the UNESCO Science Report: towards 2030[9], there is growing internationalization and multi-polarization of scientific production, research and innovation with countries and scientists collaborating across borders and disciplines to solve pressing problems and long-term challenges.

Riegler[10] posited "Interdisciplinary inquiry presupposes an open worldview to enable the researcher to transcend the confinements of a specific discipline in order to become aware of aspects that are necessary to satisfyingly solve a problem".

It is therefore timely to relook at how universities' research performance are assessed with respect to how they are able to harness the synergies created from bringing together multi-disciplinary and collaborative teams to tackle the research problems in addition to sustaining research impact.

A novel methodology has thus been designed to evaluate research multi-disciplinarity. This is complemented by indicators to gauge research impact and research collaborative-ness. Together, these three key aspects make up a new composite index for measuring, ranking and analysing Universities' research performance - World University Research Rankings. The focus is on research and the primary aim is to help the universities in the following:

(1) inform how they compare to their peers and where they stand relative to their own goals and vision;
(2) provide insights to their relative strengths and weaknesses to better capitalize on opportunities and mitigate threats; and
(3) raise awareness and promote greater research collaboration and multi-disciplinarity research amongst the global research community.

In this new World University Research Rankings 2020, an in-depth, wide-ranging study of about 12.7 million Web of Science documents across six research fields and forty-two research categories based on the OECD's schema in Web of Science's InCites was undertaken. The study was conducted across a 10-year timeframe from 2009 to 2018, where 250 top global universities were selected based on QS 2020 World University Rankings. The methodology, results and findings, including analysis by geographical representation (location and continent) and correlation between indicators and aspects, are reported in detail in this publication.

2. Methodology

Three key aspects for the World University Research Rankings have been identified, namely research multi-disciplinarity, research impact and research collaborative-ness. To quantitatively measure these three aspects, seven indicators were selected with Web of Science as the source of data as shown in Table 1 below.

Table 1: Aspects and indicators in World University Research Rankings

Aspect	Indicator	Measure
Research multi-disciplinarity	% multi-field documents	Degree of multi-disciplinary research across different fields
	% multi-category documents	Degree of multi-disciplinary research across different categories within same research field
Research impact	Category Normalised Citation Impact	Average standard of University's research impact
	% documents in Q1 journals	Proportion of University's publications in most impactful journals
	% documents in top 1%	Proportion of peaks of excellence in University's research impact
Research collaborative-ness	% industry collaboration	Degree of collaborative-ness across academia-industry boundaries
	% international collaboration	Degree of collaborative-ness across international borders

To define and measure the % of multi-field and multi-category documents, OECD's schema in Web of Science's InCites which classifies fields of sciences and technology in forty-two categories under six major fields were selected. A multi-category document was defined as a document which is classified in two or more OECD categories within the same field; and a multi-field document was defined as a document which is classified in two or more OECD fields, based on its research areas in Web of Science. The remaining five indicators are defined in InCites Indicators Handbook[11] as shown in Table 2 below and the values for respective universities can be obtained directly from InCites.

Table 2: Definition of indicators from InCites Indicators Handbook

Indicator	Definition
Category Normalised Citation Impact	The Category Normalised Citation Impact (CNCI) of a document is calculated by dividing the actual count of citing items by the expected citation rate for documents with the same document type, year of publication and subject area. When a document is assigned to more than one subject area, an average of the ratios of the actual to expected citations is used. The CNCI of a set of documents, for example, the collected works of an individual, institution, or location, is the average of the CNCI values for all the documents in the set.
% documents in Q1 journals	% of documents that appear in a journal belonging to the top Journal Impact Factor Quartile in a given year.
% documents in top 1%	The % Documents in Top 1% indicator is the top one percent most cited documents (as defined in the description of Average Percentile) in a given subject category, year and publication type divided by the total number of documents in a given set of documents, displayed as a percentage.
% industry collaboration	An industry collaborative publication is one that lists its organization type as "corporate" for one or more of the co-author's affiliations. The % of Industry Collaborations is the number of industry collaborative publications for an entity (as described above) divided by the total number of documents for the same entity represented as a percentage.
% international collaboration	The International Collaborations indicator shows the number of publications that have been found with at least two different countries among the affiliations of the co-authors. The % of International Collaborations is the number of International Collaborations for an entity (as described above) divided by the total number of documents for the same entity represented as a percentage.

For each university, its values for these seven indicators are normalised and aggregated for respective rankings (overall, research multi-disciplinarity, research impact, research collaborative-ness). To normalise for aggregation, the top performing university in an indicator is assigned a score of 100% with the remaining universities pro-rated accordingly based on their indicator's value relative to the top performer's. To illustrate, an example is shown in Table 3 below.

Table 3: Illustrated example of computation of normalised score

University	CNCI	Normalised score for CNCI
University with highest CNCI value	3	100%
University ABC	1.5	1.5/3 x 100% = 50%

To scope this, the top 250 universities in QS 2020 World University Rankings were selected; and their research performance evaluated over a 10-year timeframe from 2009 to 2018 based on the following formulae.

Research multi-disciplinarity score = Mean of M1 and M2 normalised to 100% based on top performing university where
M1 = % multi-field publications normalised to 100% based on top performing university
M2 = % multi-category publications normalised to 100% based on top performing university

Research impact score = mean of I1, I2 and I3 normalised to 100% based on top performing university where
I1 = CNCI normalised to 100% based on top performing university
I2 = % docs in top 1% docs normalised to 100% based on top performing university
I3 = % docs in Q1 journals normalised to 100% based on top performing university

Research collaborative-ness score = mean of C1 and C2 normalised to 100% based on top performing university where
C1 = % international collaboration normalised to 100% based on top performing university
C2 = % industry collaboration normalised to 100% based on top performing university

Overall research assessment score = Mean of M1, M2, I1, I2, I3, C1 and C2 normalised to 100% based on top performing university

Universities are also classified by (full time equivalent) size of the degree-seeking student body and age based on foundation years12 as shown in Tables 4 and 5 below, with QS rankings and classification data extracted from QS World University Rankings® 2020 report[13].

Table 4: Size classification

Size classification	Students
XL	$\geq 30,000$
L	$\geq 12,000$ and $< 30,000$
M	$\geq 5,000$ and $< 12,000$
S	$< 5,000$

Table 5: Age classification

Age classification	Age from foundation
V	≥ 100 years old
IV	≥ 50 and < 100 years old
III	≥ 25 and < 50 years old
II	≥ 10 and < 25 years old
I	< 10 years old

3. Results and analysis

All 250 universities are ranked based on their overall research assessment score as shown in Table 6 below.

Table 6: Global and regional rankings based on overall research assessment score

University	Location	Region	Overall score	Global rank	Regional rank	QS rank	Size	Age
Massachusetts Institute of Technology (MIT)	USA	N America	100.0%	1	1	1	M	V
Technical University of Denmark	Denmark	Europe	95.2%	2	1	112	M	V
Ecole Polytechnique Federale de Lausanne	Switzerland	Europe	94.1%	3	2	18=	M	V
Carnegie Mellon University	USA	N America	92.5%	4	2	48	L	V
Eindhoven University of Technology	Netherlands	Europe	91.3%	5	3	102	M	IV
CentraleSupelec	France	Europe	89.1%	6	4	139	S	I^
Telecom ParisTech	France	Europe	89.1%	7	5	224=	S	V*
Wageningen University & Research	Netherlands	Europe	88.9%	8	6	125=	L	V
ETH Zurich	Switzerland	Europe	88.2%	9	7	6	L	V
Nanyang Technological University	Singapore	Asia	88.2%	10	1	11=	L	III
Stanford University	USA	N America	88.1%	11	3	2	L	V
Chalmers University of Technology	Sweden	Europe	87.5%	12	8	125=	M	V
Georgia Institute of Technology	USA	N America	85.5%	13	4	72=	L	V
Royal Institute of Technology	Sweden	Europe	85.3%	14	9	98=	L	V
University of California Berkeley	USA	N America	85.3%	15	5	28	XL	V
Harvard University	USA	N America	84.8%	16	6	3	L	V
Princeton University	USA	N America	84.7%	17	7	13	M	V
California Institute of Technology	USA	N America	84.3%	18	8	5	S	V
KU Leuven	Belgium	Europe	84.3%	19	10	80	XL	V
Delft University of Technology	Netherlands	Europe	84.2%	20	11	50	L	V
Ecole des Ponts ParisTech	France	Europe	84.2%	21	12	250	S	V
University of California Santa Barbara	USA	N America	83.5%	22	9	135	L	V

Table 6: (*Continued*)

University	Location	Region	Overall score	Global rank	Regional rank	QS rank	Size	Age
Aalto University	Finland	Europe	82.9%	23	13	134	L	I^
King Abdulaziz University	Saudi Arabia	Asia	82.5%	24	2	186=	XL	IV
Imperial College London	UK	Europe	82.2%	25	14	9	L	V
University of California San Diego	USA	N America	81.9%	26	10	45	XL	IV
Hong Kong University of Science & Technology	Hong Kong	Asia	81.8%	27	3	32	M	III
University of Basel	Switzerland	Europe	81.7%	28	15	151	M	V
Rice University	USA	N America	81.3%	29	11	85	M	V
Pohang University of Science & Technology (POSTECH)	Korea	Asia	80.9%	30	4	87	S	III
University of Twente	Netherlands	Europe	80.7%	31	16	186=	M	IV
University of Copenhagen	Denmark	Europe	80.5%	32	17	81=	XL	V
University of Cambridge	UK	Europe	80.1%	33	18	7	L	V
Duke University	USA	N America	79.9%	34	12	25=	L	V
Queen Mary University London	UK	Europe	79.3%	35	19	126	L	V
University of Washington Seattle	USA	N America	79.1%	36	13	68	XL	V
Erasmus University Rotterdam	Netherlands	Europe	78.9%	37	20	183	L	V
Polytechnic University of Milan	Italy	Europe	78.9%	38	21	149	XL	V
University of Chicago	USA	N America	78.8%	39	14	10	L	V
National University of Singapore	Singapore	Asia	78.6%	40	5	11=	XL	V
Technical University of Munich	Germany	Europe	78.5%	41	22	55	XL	V
Scuola Normale Superiore di Pisa	Italy	Europe	78.3%	42	23	204	S	V*
University of Oxford	UK	Europe	78.3%	43	24	4	L	V
University of Maryland College Park	USA	N America	78.2%	44	15	136	XL	V
Utrecht University	Netherlands	Europe	78.1%	45	25	120=	XL	V
Universite Catholique Louvain	Belgium	Europe	77.9%	46	26	167=	L	V

Table 6: (*Continued*)

University	Location	Region	Overall score	Global rank	Regional rank	QS rank	Size	Age
Uppsala University	Sweden	Europe	77.6%	47	27	116=	L	V
Radboud University Nijmegen	Netherlands	Europe	77.3%	48	28	217=	L	IV
Vrije Universiteit Amsterdam	Netherlands	Europe	77.3%	49	29	219=	L	V
Columbia University	USA	N America	77.2%	50	16	18=	L	V
Scuola Superiore Sant'Anna	Italy	Europe	77.1%	51	30	177=	S	III
University of Amsterdam	Netherlands	Europe	76.9%	52	31	64	L	V
University of Zurich	Switzerland	Europe	76.9%	53	32	76	L	V
Vienna University of Technology	Austria	Europe	76.9%	54	33	192=	M	V
University of California Los Angeles	USA	N America	76.8%	55	17	35=	XL	V
Karlsruhe Institute of Technology	Germany	Europe	76.8%	56	34	124	L	IV
Korea Advanced Institute of Science & Technology (KAIST)	Korea	Asia	76.7%	57	6	41	M	III
University of Edinburgh	UK	Europe	76.7%	58	35	20	XL	V
Northwestern University	USA	N America	76.6%	59	18	31	L	V
University College London	UK	Europe	76.6%	60	36	8	XL	V
University of Texas Austin	USA	N America	76.5%	61	19	65	XL	V
PSL Research University Paris (ComUE)	France	Europe	76.5%	62	37	53	L	I^
Yale University	USA	N America	76.4%	63	20	17	L	V
University of Technology Sydney	Australia	Oceania	76.3%	64	1	140=	L	III
Johns Hopkins University	USA	N America	76.3%	65	21	24	L	V
Ecole Polytechnique	France	Europe	76.3%	66	38	60=	S	V
University of Groningen	Netherlands	Europe	76.2%	67	39	114	L	V
Stockholm University	Sweden	Europe	76.2%	68	40	191	L	V
University of Antwerp	Belgium	Europe	76.1%	69	41	223	L	III
Washington University (WUSTL)	USA	N America	76.1%	70	22	108=	L	V

Table 6: (*Continued*)

University	Location	Region	Overall score	Global rank	Regional rank	QS rank	Size	Age
University of Colorado Boulder	USA	N America	76.1%	71	23	206	XL	V
University of Pennsylvania	USA	N America	76.1%	72	24	15	L	V
City University of Hong Kong	Hong Kong	Asia	75.9%	73	7	52	M	III
Technical University of Berlin	Germany	Europe	75.9%	74	42	147	L	V
University of Southampton	UK	Europe	75.8%	75	43	97	L	V
Lund University	Sweden	Europe	75.7%	76	44	92	L	V
Cornell University	USA	N America	75.6%	77	25	14	L	V
Leiden University	Netherlands	Europe	75.0%	78	45	118	XL	V
RWTH Aachen University	Germany	Europe	74.9%	79	46	138	XL	V
University of Southern California	USA	N America	74.8%	80	26	129	XL	V
Boston University	USA	N America	74.8%	81	27	98=	L	V
University of Geneva	Switzerland	Europe	74.8%	82	47	110	L	V
Dresden University of Technology	Germany	Europe	74.8%	83	48	179=	XL	V
University of Toronto	Canada	N America	74.7%	84	28	29=	XL	V
Aarhus University	Denmark	Europe	74.7%	85	49	145	L	IV
Vanderbilt University	USA	N America	74.6%	86	29	200=	L	V
University of North Carolina Chapel Hill	USA	N America	74.5%	87	30	90	XL	V
Kings College London	UK	Europe	74.4%	88	50	33=	L	V
University of Illinois Urbana-Champaign	USA	N America	74.4%	89	31	75	XL	V
University of Barcelona	Spain	Europe	74.4%	90	51	165=	XL	V
Hong Kong Polytechnic University	Hong Kong	Asia	74.3%	91	8	91	L	III
Ghent University	Belgium	Europe	74.3%	92	52	130=	XL	V
McMaster University	Canada	N America	74.2%	93	32	140=	L	V
McGill University	Canada	N America	74.0%	94	33	35=	L	V

Table 6: (*Continued*)

University	Location	Region	Overall score	Global rank	Regional rank	QS rank	Size	Age
University of Bristol	UK	Europe	73.9%	95	53	49	L	V
Humboldt University of Berlin	Germany	Europe	73.9%	96	54	120=	XL	V
University of Waterloo	Canada	N America	73.8%	97	34	173=	XL	IV
University of Oslo	Norway	Europe	73.7%	98	55	119	L	V
University of Lausanne	Switzerland	Europe	73.7%	99	56	153	L	V
Ruprecht Karls University Heidelberg	Germany	Europe	73.6%	100	57	66	L	V
University of Sussex	UK	Europe	73.6%	101	58	246	L	IV
Lancaster University	UK	Europe	73.5%	102	59	128	L	IV
University of California Irvine	USA	N America	73.4%	103	35	219=	XL	IV
University of Glasgow	UK	Europe	73.4%	104	60	67	L	V
Maastricht University	Netherlands	Europe	73.1%	105	61	239=	L	III
Vrije Universiteit Brussel	Belgium	Europe	73.0%	106	62	195	M	V
Chinese University of Hong Kong	Hong Kong	Asia	72.9%	107	9	46	L	IV
Tsinghua University	China (Mainland)	Asia	72.9%	108	10	16	XL	V
University of Manchester	UK	Europe	72.9%	109	63	27	XL	V
Sorbonne Universite	France	Europe	72.8%	110	64	77	XL	V*
Arizona State University	USA	N America	72.7%	111	36	215=	XL	V
University of Michigan	USA	N America	72.6%	112	37	21	XL	V
University of British Columbia	Canada	N America	72.5%	113	38	51	XL	V
University of Helsinki	Finland	Europe	72.4%	114	65	107	L	V
University of Munich	Germany	Europe	72.1%	115	66	63	XL	V
Newcastle University - UK	UK	Europe	72.0%	116	67	146	L	IV
University of Wollongong	Australia	Oceania	72.0%	117	2	212=	L	III
Free University of Berlin	Germany	Europe	71.9%	118	68	130=	XL	IV

Table 6: (*Continued*)

University	Location	Region	Overall score	Global rank	Regional rank	QS rank	Size	Age
University of Bath	UK	Europe	71.8%	119	69	172	L	IV
Eberhard Karls University of Tubingen	Germany	Europe	71.7%	120	70	169=	L	V
University of Bergen	Norway	Europe	71.5%	121	71	163=	M	IV
Emory University	USA	N America	71.4%	122	39	156=	L	V
New York University	USA	N America	71.3%	123	40	39	XL	V
Cardiff University	UK	Europe	71.2%	124	72	154=	L	V
University of Liverpool	UK	Europe	71.2%	125	73	181	L	V
University of Aberdeen	UK	Europe	71.1%	126	74	194	L	V
University of Bologna	Italy	Europe	71.1%	127	75	177=	XL	V
University of Cape Town	South Africa	Africa	71.1%	128	1	198=	L	V
Purdue University	USA	N America	71.0%	129	41	111	XL	V
University of Navarra	Spain	Europe	70.9%	130	76	245	M	IV
University of Sheffield	UK	Europe	70.9%	131	77	78=	L	V
University of Wisconsin Madison	USA	N America	70.7%	132	42	56	XL	V
University of Bern	Switzerland	Europe	70.6%	133	78	123	M	V
University of Bonn	Germany	Europe	70.6%	134	79	243	XL	V
Royal Melbourne Institute of Technology (RMIT)	Australia	Oceania	70.5%	135	3	238	XL	III
National Tsing Hua University	Taiwan	Asia	70.3%	136	11	173=	L	IV
University of Rochester	USA	N America	70.3%	137	43	171	M	V
University of Exeter	UK	Europe	70.3%	138	80	163=	L	IV
University of Pittsburgh	USA	N America	70.2%	139	44	140=	L	V
University of Birmingham	UK	Europe	70.0%	140	81	81=	L	V
University of Adelaide	Australia	Oceania	69.9%	141	4	106	L	V
University of Minnesota Twin Cities	USA	N America	69.9%	142	45	156=	XL	V

Table 6: (*Continued*)

University	Location	Region	Overall score	Global rank	Regional rank	QS rank	Size	Age
University of Leeds	UK	Europe	69.9%	143	82	93=	XL	V
Autonomous University of Barcelona	Spain	Europe	69.8%	144	83	188	XL	IV
Ohio State University	USA	N America	69.8%	145	46	101	XL	V
University of New South Wales Sydney	Australia	Oceania	69.7%	146	5	43	XL	IV
University of Nottingham	UK	Europe	69.6%	147	84	96	XL	V
University of York - UK	UK	Europe	69.6%	148	85	148	L	IV
University of California Davis	USA	N America	69.6%	149	47	104=	XL	V
University of Montreal	Canada	N America	69.6%	150	48	137	XL	V
University of Freiburg	Germany	Europe	69.5%	151	86	169=	L	V
University of Queensland	Australia	Oceania	69.4%	152	6	47	XL	V
Ecole Normale Superieure de Lyon (ENS LYON)	France	Europe	69.3%	153	87	160=	S	V
Loughborough University	UK	Europe	69.3%	154	88	222	L	IV
University of Science & Technology of China	China (Mainland)	Asia	69.0%	155	12	89	L	IV
University of Padua	Italy	Europe	69.0%	156	89	234=	XL	V
University of Melbourne	Australia	Oceania	69.0%	157	7	38	XL	V
University of Reading	UK	Europe	68.9%	158	90	205	L	V
University of Auckland	New Zealand	Oceania	68.8%	159	8	83=	L	V
Queens University Belfast	UK	Europe	68.8%	160	91	173=	L	V
University of Leicester	UK	Europe	68.7%	161	92	239=	L	IV
University of Warwick	UK	Europe	68.7%	162	93	62	L	IV
Brown University	USA	N America	68.5%	163	49	57	M	V
Sungkyunkwan University (SKKU)	Korea	Asia	68.5%	164	13	95	L	V
Penn State University	USA	N America	68.5%	165	50	93=	XL	V
University of Hamburg	Germany	Europe	68.5%	166	94	227=	XL	V

Table 6: (*Continued*)

University	Location	Region	Overall score	Global rank	Regional rank	QS rank	Size	Age
University of Vienna	Austria	Europe	68.3%	167	95	154=	XL	V
University of Alberta	Canada	N America	68.3%	168	51	113	XL	V
University of Calgary	Canada	N America	68.2%	169	52	233	XL	IV
University of Sydney	Australia	Oceania	68.1%	170	9	42	XL	V
University of Gottingen	Germany	Europe	68.1%	171	96	197	XL	V
Australian National University	Australia	Oceania	68.0%	172	10	29=	L	IV
Dartmouth College	USA	N America	67.9%	173	53	207=	M	V
University of Hong Kong	Hong Kong	Asia	67.8%	174	14	25=	L	V
University of Canterbury	New Zealand	Oceania	67.7%	175	11	227=	L	V
Curtin University	Australia	Oceania	67.7%	176	12	230	L	III
University of the Andes Colombia	Colombia	S America	67.6%	177	1	234=	L	IV
University of Western Australia	Australia	Oceania	67.5%	178	13	86	L	V
Georgetown University	USA	N America	67.2%	179	54	226	L	V
London School Economics & Political Science	UK	Europe	67.1%	180	97	44	M	V
Trinity College Dublin	Ireland	Europe	67.1%	181	98	108=	L	V
Case Western Reserve University	USA	N America	66.9%	182	55	179=	M	V
Monash University	Australia	Oceania	66.9%	183	14	58=	XL	IV
University of Virginia	USA	N America	66.6%	184	56	198=	L	V
University College Dublin	Ireland	Europe	66.3%	185	99	185	L	V
Michigan State University	USA	N America	66.1%	186	57	144	XL	V
Autonomous University of Madrid	Spain	Europe	66.0%	187	100	192=	L	IV
Texas A&M University College Station	USA	N America	65.9%	188	58	189=	XL	V
Durham University	UK	Europe	65.8%	189	101	78=	L	V
Queensland University of Technology (QUT)	Australia	Oceania	65.7%	190	15	224=	XL	III

Table 6: (*Continued*)

University	Location	Region	Overall score	Global rank	Regional rank	QS rank	Size	Age
Hebrew University of Jerusalem	Israel	Asia	65.6%	191	15	162	L	V
University of St Andrews	UK	Europe	65.4%	192	102	100	M	V
University of Notre Dame	USA	N America	65.3%	193	59	210	L	V
National Chiao Tung University	Taiwan	Asia	65.0%	194	16	227=	M	V
University of Florida	USA	N America	64.5%	195	60	167=	XL	V
National Taiwan University	Taiwan	Asia	64.4%	196	17	69	XL	IV
Tel Aviv University	Israel	Asia	64.3%	197	18	219=	L	IV
Tokyo Institute of Technology	Japan	Asia	64.1%	198	19	58=	M	V
Seoul National University (SNU)	Korea	Asia	64.0%	199	20	37	L	IV
Queens University - Canada	Canada	N America	63.8%	200	61	239=	L	V
Peking University	China (Mainland)	Asia	63.5%	201	21	22=	XL	V
Macquarie University	Australia	Oceania	63.4%	202	16	237	L	IV
King Fahd University of Petroleum & Minerals	Saudi Arabia	Asia	63.4%	203	22	200=	M	IV
Sapienza University Rome	Italy	Europe	63.2%	204	103	203	XL	V
Waseda University	Japan	Asia	63.0%	205	23	196	XL	V
American University of Beirut	Lebanon	Asia	63.0%	206	24	244	M	V
University of Otago	New Zealand	Oceania	62.7%	207	17	176	L	V
Western University (University of Western Ontario)	Canada	N America	62.6%	208	62	211	L	V
Tecnologico de Monterrey	Mexico	N America	62.4%	209	63	158	L	IV
Hanyang University	Korea	Asia	62.2%	210	25	150	L	IV
Universiti Teknologi Malaysia	Malaysia	Asia	61.5%	211	26	217=	L	III
University of Newcastle	Australia	Oceania	61.5%	212	18	207=	L	IV
University of Tokyo	Japan	Asia	61.3%	213	27	22=	L	V
Victoria University Wellington	New Zealand	Oceania	61.2%	214	19	215=	L	V

Table 6: (*Continued*)

University	Location	Region	Overall score	Global rank	Regional rank	QS rank	Size	Age
Shanghai Jiao Tong University	China (Mainland)	Asia	61.2%	215	28	60=	XL	V
Korea University	Korea	Asia	61.0%	216	29	83=	L	V
National Cheng Kung University	Taiwan	Asia	60.7%	217	30	225	L	IV
Indian Institute of Technology (IIT) - Bombay	India	Asia	59.6%	218	31	152	M	IV
Pontificia Universidad Catolica de Chile	Chile	S America	59.6%	219	2	127	L	V
University of Illinois Chicago	USA	N America	59.4%	220	64	231=	L	V
Zhejiang University	China (Mainland)	Asia	59.3%	221	32	54	XL	V
Fudan University	China (Mainland)	Asia	59.3%	222	33	40	XL	V
Tohoku University	Japan	Asia	59.1%	223	34	82	L	V
Nanjing University	China (Mainland)	Asia	59.0%	224	35	120=	XL	V
Yonsei University	Korea	Asia	58.5%	225	36	104=	L	V
Kyoto University	Japan	Asia	58.4%	226	37	33=	L	V
Universiti Malaya	Malaysia	Asia	58.3%	227	38	70	L	V
Indian Institute of Science (IISC) - Bangalore	India	Asia	58.2%	228	39	184	S	V
Chulalongkorn University	Thailand	Asia	57.9%	229	40	247=	XL	V
Osaka University	Japan	Asia	57.2%	230	41	71	L	IV
Complutense University of Madrid	Spain	Europe	57.1%	231	104	212=	XL	V
Institut d'Etudes Politiques Paris (SciencePo)	France	Europe	56.9%	232	105	242	M	V
Indian Institute of Technology (IIT) - Delhi	India	Asia	56.8%	233	42	182	M	IV
Kyushu University	Japan	Asia	55.7%	234	43	132=	L	V
Universiti Sains Malaysia	Malaysia	Asia	55.6%	235	44	165=	L	IV

Table 6: (*Continued*)

University	Location	Region	Overall score	Global rank	Regional rank	QS rank	Size	Age
Universidad de Chile	Chile	S America	55.4%	236	3	189=	XL	V
Nagoya University	Japan	Asia	55.0%	237	45	115	L	V
University of Buenos Aires	Argentina	S America	54.5%	238	4	74	XL	V
Universiti Putra Malaysia	Malaysia	Asia	54.3%	239	46	159	L	III
Keio University	Japan	Asia	53.8%	240	47	200=	XL	V
Kyung Hee University	Korea	Asia	52.7%	241	48	247=	L	IV
Hokkaido University	Japan	Asia	52.5%	242	49	132=	L	V
Universidade Estadual de Campinas	Brazil	S America	52.1%	243	5	214	L	IV
Universidade de Sao Paulo	Brazil	S America	49.1%	244	6	116=	XL	IV
Universiti Kebangsaan Malaysia	Malaysia	Asia	48.5%	245	50	160=	L	III
Novosibirsk State University	Russia	Europe	47.9%	246	106	231=	M	IV
Universidad Nacional Autonoma de Mexico	Mexico	N America	47.5%	247	65	103	XL	V
Saint Petersburg State University	Russia	Europe	45.6%	248	107	234=	L	V
Lomonosov Moscow State University	Russia	Europe	41.9%	249	108	84	XL	V
Al-Farabi Kazakh National University	Kazakhstan	Asia	41.8%	250	51	207=	L	IV

*Data not available in QS report hence was obtained separately based on information in the University's website[14-16]

^Based on year of merger

3.1. *Profile of Universities*

The breakdown of all 250 universities in this study by size, age and region are shown in Figures 1, 2 and 3 below.

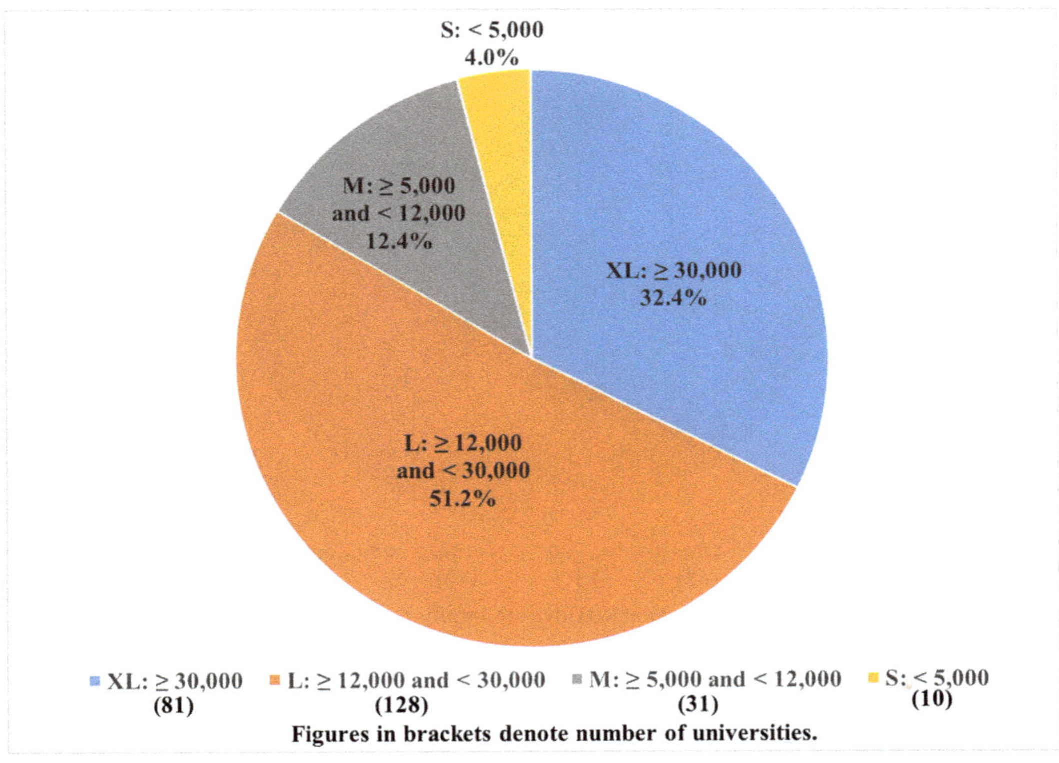

Figure 1: Size of all 250 universities

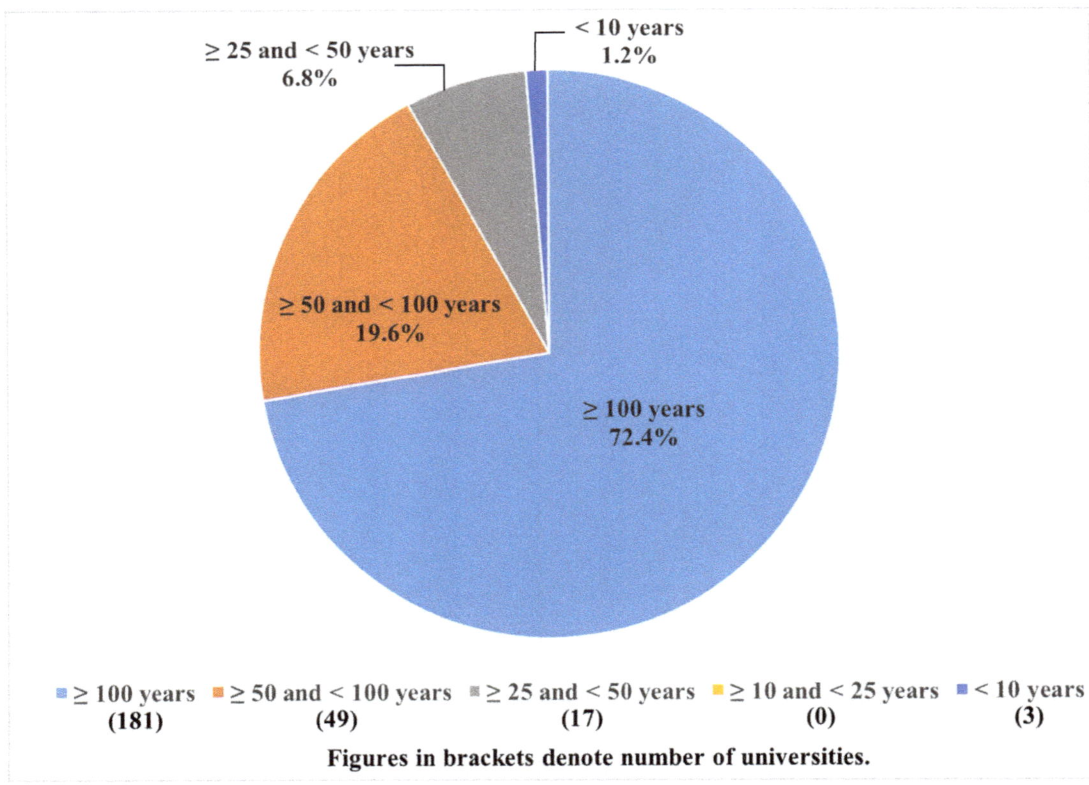

Figure 2: Age of all 250 universities

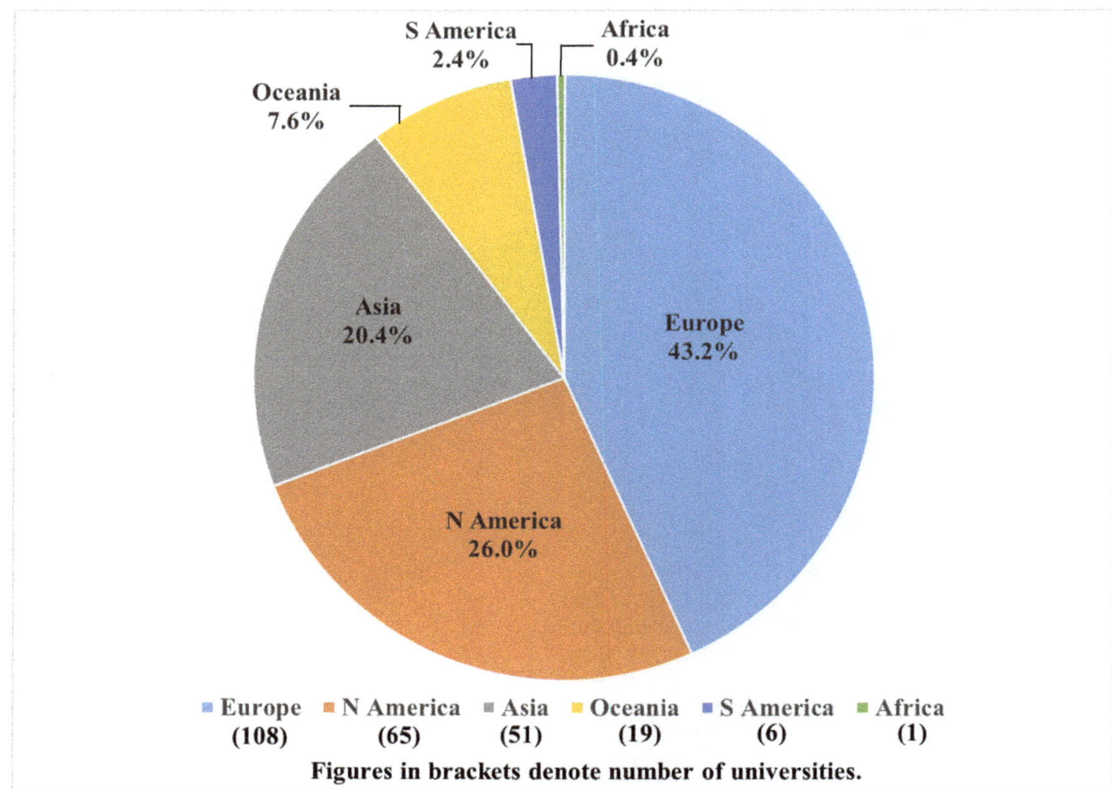

Figure 3: Region of all 250 universities

Among all the universities in this study, 83.6% are of large to extra-large in size and 16.4% are of small to medium in size; 92% are at least 50 years old and 8% are younger than 50 years; 69.2% are from Europe and North America and 30.8% are from the rest of the world.

The breakdown of the top 50 universities in this study by size, age and region are shown in Figures 4, 5 and 6 below.

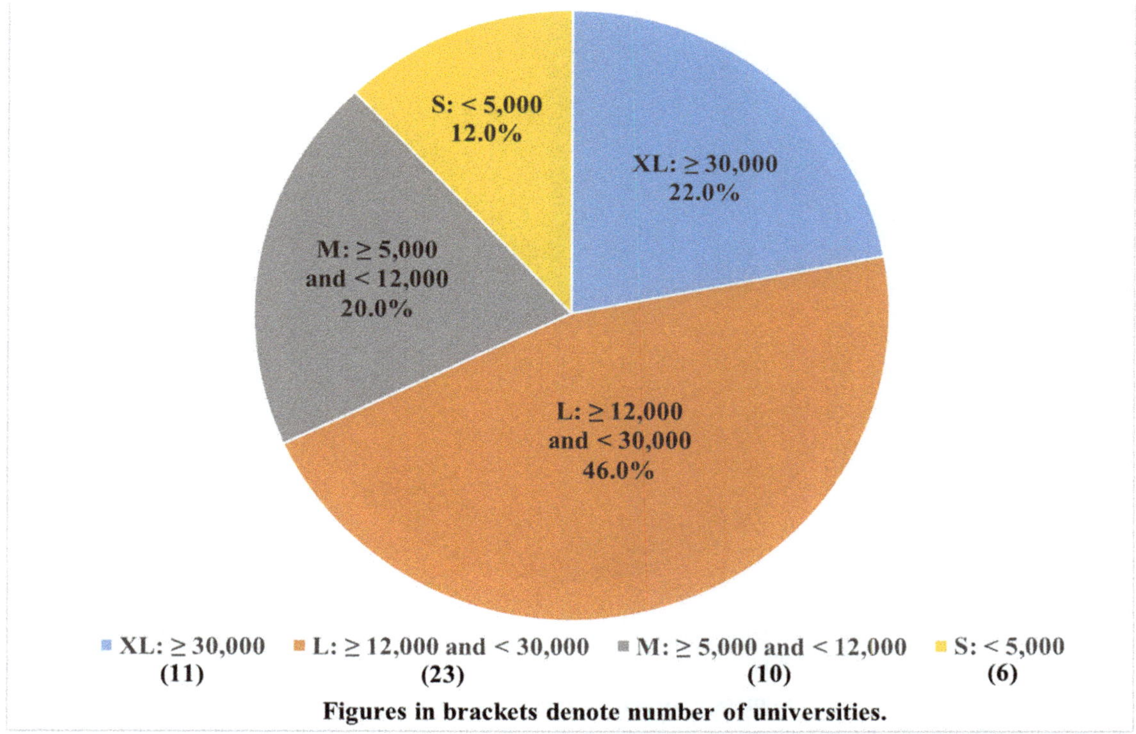

Figure 4: Size of top 50 universities

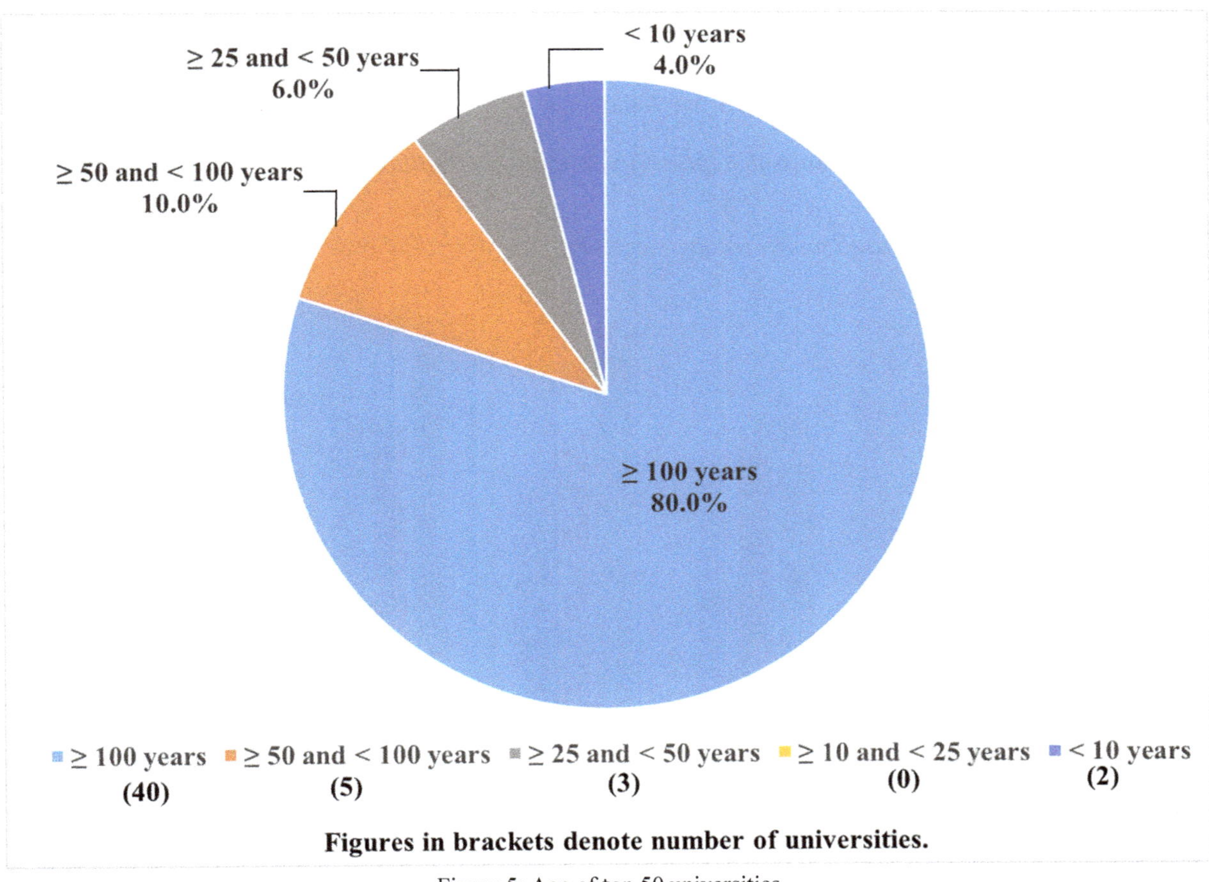

Figure 5: Age of top 50 universities

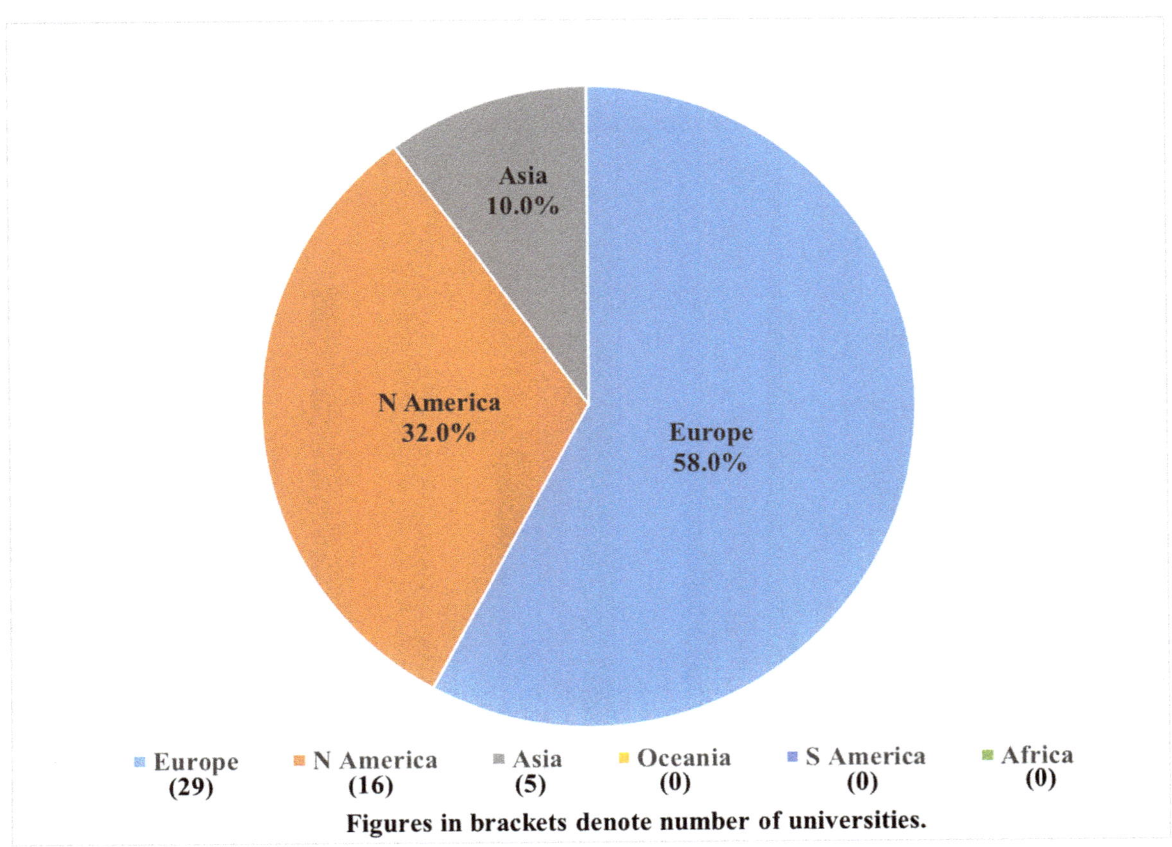

Figure 6: Region of top 50 universities

Among the top 50 universities in this study, 68% are of large to extra-large in size and 32% are of small to medium in size; 90% are at least 50 years old and 10% are less than 50 years old; 90% are from Europe and North America and 10% are from Asia.

3.2. *Comparison of research multi-disciplinarity, impact and collaborative-ness scores by size, age, region and location*

3.2.1. *Comparison of research multi-disciplinarity, impact and collaborative-ness scores by size*

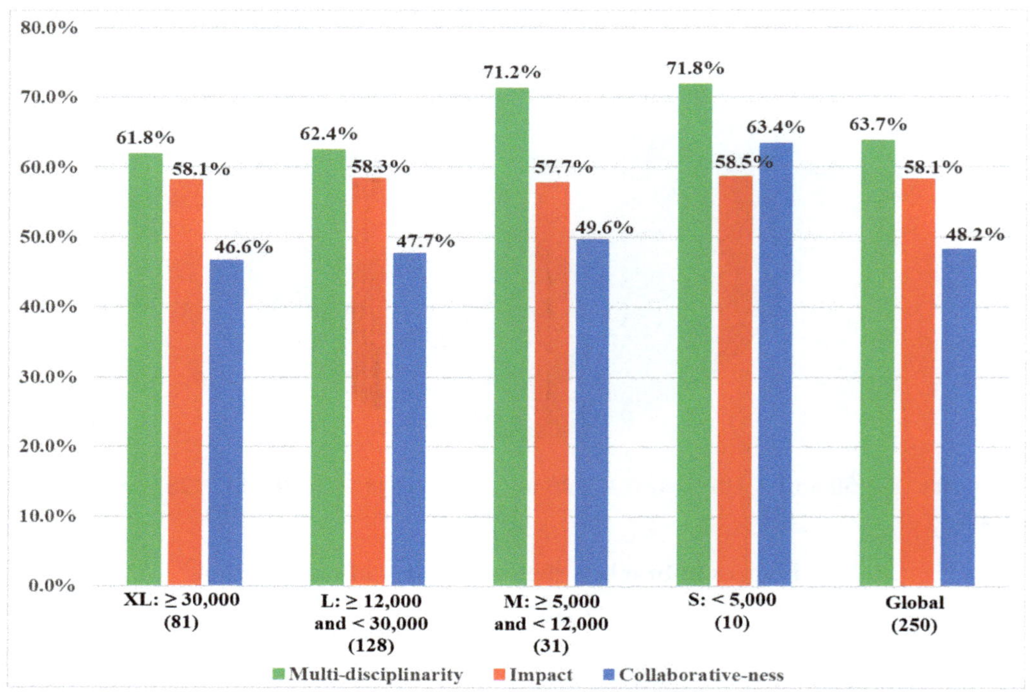

Figure 7: Comparison of multi-disciplinarity, impact and collaborative-ness scores by size
Figures in brackets denote number of universities.

3.2.2. *Comparison of research multi-disciplinarity, impact and collaborative-ness scores by age*

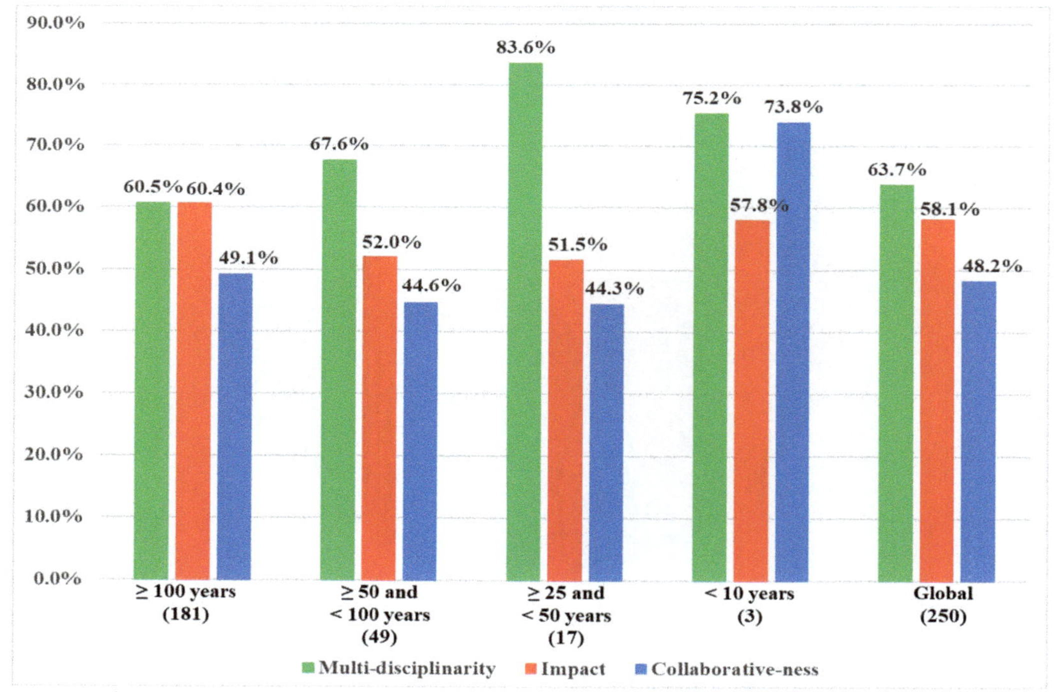

Figure 8: Comparison of multi-disciplinarity, impact and collaborative-ness scores by age
Figures in brackets denote number of universities.

3.2.3. *Comparison of research multi-disciplinarity, impact and collaborative-ness scores by region*

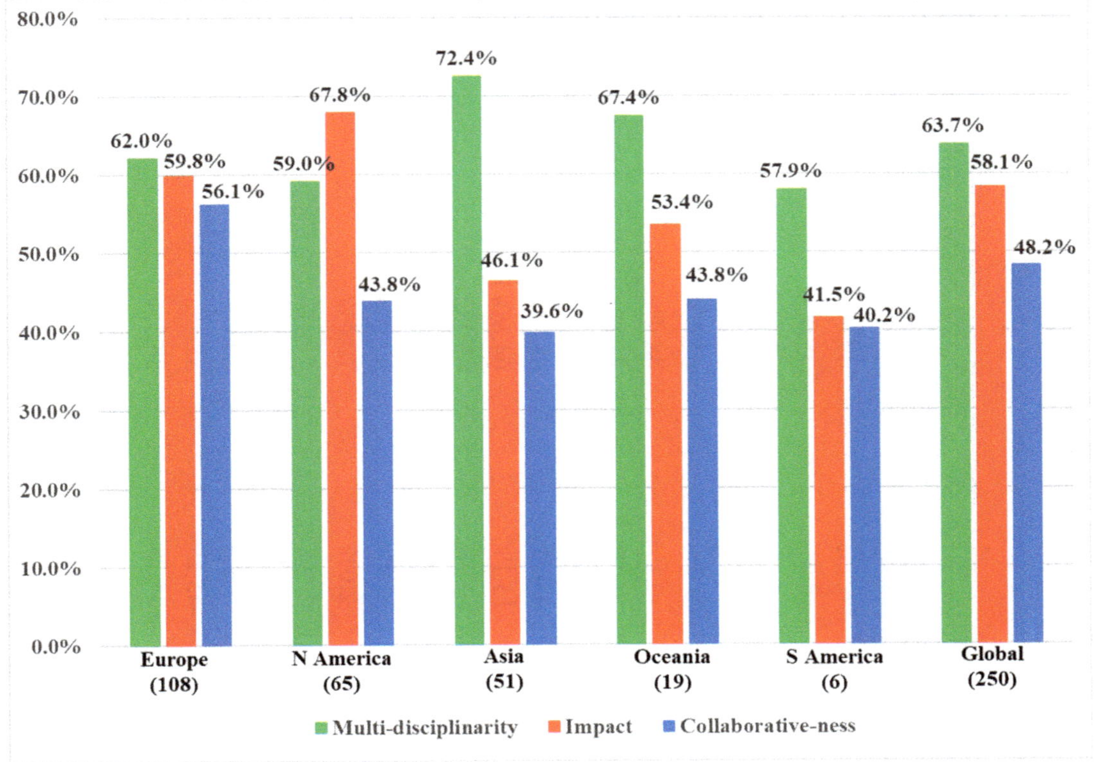

Figure 9: Comparison of multi-disciplinarity, impact and collaborative-ness scores by region
Africa is excluded as there is only one representative. Figures in brackets denote number of universities.

3.2.4 Comparison of research multi-disciplinarity, impact and collaborative-ness scores by location

Table 7: Normalised location averages for research multi-disciplinarity, impact and collaborative-ness
(figures in brackets denote number of universities represented in this study)

Location	Multi-disciplinarity Norm. score	Multi-disciplinarity % var. vs global	Impact Norm. score	Impact % var. vs global	Collaborative-ness Norm. score	Collaborative-ness % var. vs global
USA (53)	58.3%	-8.5%	70.9%	21.9%	43.0%	-10.8%
UK (32)	58.8%	-7.7%	62.0%	6.7%	49.7%	3.0%
China (16)	76.4%	19.9%	53.0%	-8.9%	35.0%	-27.4%
Australia (15)	68.1%	7.0%	54.4%	-6.4%	43.2%	-10.4%
Germany (14)	57.7%	-9.4%	59.4%	2.1%	57.7%	19.6%
Netherlands (12)	67.1%	5.4%	63.9%	9.9%	61.4%	27.4%
Canada (10)	60.7%	-4.6%	58.1%	-0.1%	48.9%	1.4%
Japan (10)	59.5%	-6.6%	42.5%	-26.9%	41.8%	-13.4%
France (8)	69.9%	9.7%	55.2%	-5.1%	65.5%	35.9%
Korea (8)	72.2%	13.4%	46.6%	-19.8%	45.1%	-6.5%
Switzerland (7)	58.3%	-8.5%	66.9%	15.1%	64.8%	34.3%
Italy (6)	69.6%	9.3%	58.0%	-0.3%	49.1%	1.8%
Belgium (5)	64.1%	0.7%	60.6%	4.2%	62.2%	29.0%
Spain (5)	59.7%	-6.3%	55.5%	-4.6%	47.0%	-2.4%
Sweden (5)	67.8%	6.5%	59.7%	2.6%	70.4%	46.1%
Malaysia (5)	84.1%	32.1%	31.7%	-45.6%	31.3%	-35.1%
New Zealand (4)	64.4%	1.2%	49.3%	-15.2%	46.1%	-4.3%
Denmark (3)	70.6%	10.9%	63.0%	8.4%	70.7%	46.7%
Russia (3)	49.1%	-22.8%	31.9%	-45.2%	32.1%	-33.5%
India (3)	83.9%	31.7%	37.7%	-35.2%	28.3%	-41.4%

Table 7: (*Continued*)

Location	Multi-disciplinarity Norm. score	Multi-disciplinarity % var. vs global	Impact Norm. score	Impact % var. vs global	Collaborative-ness Norm. score	Collaborative-ness % var. vs global
Singapore (2)	82.4%	29.5%	63.9%	9.9%	57.8%	20.0%
Chile (2)	51.6%	-19.0%	43.4%	-25.3%	45.9%	-4.8%
Brazil (2)	62.4%	-2.0%	35.3%	-39.3%	29.8%	-38.2%
Finland (2)	70.1%	10.1%	60.0%	3.2%	59.1%	22.5%
Norway (2)	53.8%	-15.4%	59.9%	3.0%	59.4%	23.1%
Israel (2)	57.9%	-9.1%	53.1%	-8.6%	44.9%	-6.9%
Saudi Arabia (2)	75.8%	19.1%	51.7%	-11.1%	54.7%	13.4%
Ireland (2)	57.1%	-10.3%	54.9%	-5.6%	47.6%	-1.4%
Mexico (2)	70.9%	11.4%	33.9%	-41.8%	37.6%	-22.0%
Austria (2)	71.7%	12.5%	52.1%	-10.5%	57.0%	18.3%
South Africa (1)	58.7%	-7.8%	57.5%	-1.1%	54.6%	13.2%
Argentina (1)	55.8%	-12.4%	39.0%	-32.8%	41.1%	-14.8%
Thailand (1)	70.4%	10.6%	36.6%	-37.1%	42.0%	-12.9%
Lebanon (1)	56.8%	-10.7%	45.9%	-21.1%	53.1%	10.2%
Kazakhstan (1)	68.5%	7.5%	16.1%	-72.4%	32.6%	-32.4%
Colombia (1)	63.8%	0.2%	52.4%	-9.9%	48.7%	1.1%
Global (250)	63.7%	0.0%	58.1%	0.0%	48.2%	0.0%

3.3. *Drill-down analysis into performance across seven individual indicators by region and location*

3.3.1. *Drill-down analysis into performance across seven individual indicators by region*

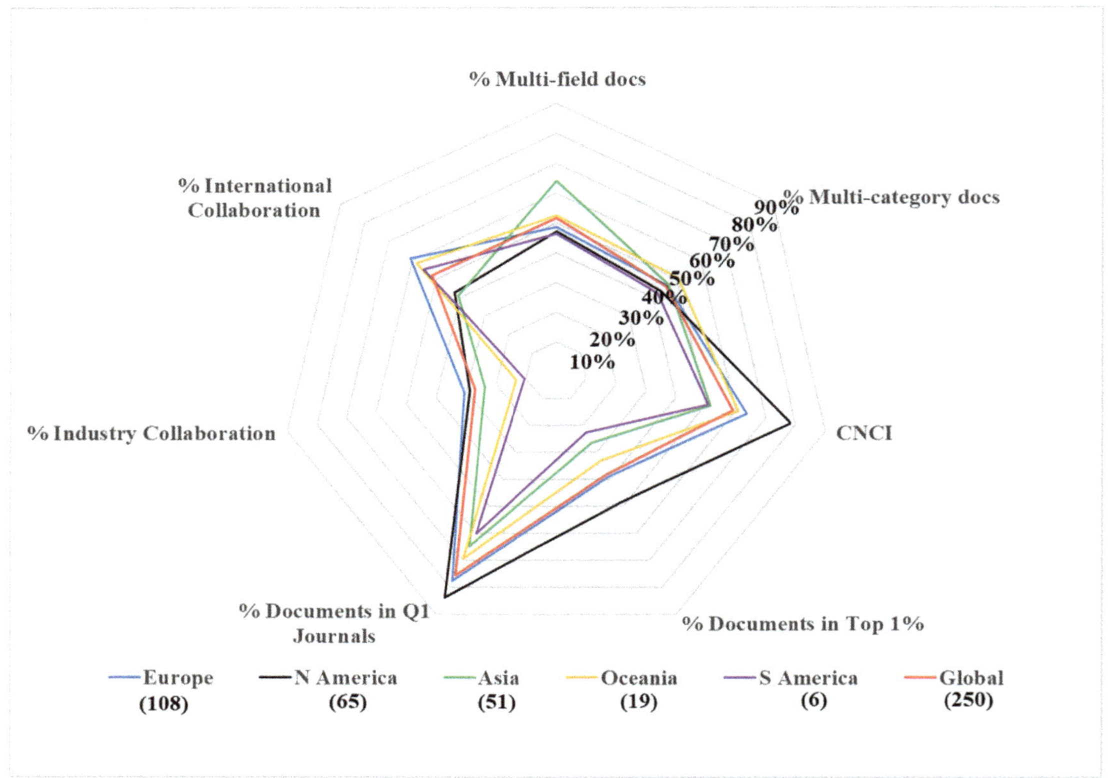

Figure 10: Comparison of region averages across all seven indicators
Africa is excluded as there is only one representative. Figures in brackets denote number of universities.

Comparing regional averages in the first aspect of the World University Research Rankings: research multi-disciplinarity, Asian universities and Oceanian universities lead the pack respectively for multi-field and multi-category research. Against the backdrop of increasing complexity, sophistication and integration of technology, multi-disciplinary research in the fields of "Engineering and Technology" and "Natural sciences" has grown significantly and STEM-focused universities from Asia and Oceania have put strong emphasis on these fields as their societies and economies evolved in tandem.

In the second aspect of the World University Research Rankings: research impact, North American universities dominate the field across all three indicators, namely CNCI, % documents in top 1% and % documents in Q1 journals. The attractiveness of North American universities in drawing global talent coupled with their relatively deep pool of resources and longer runway in establishing a track record (about three quarters of those in this study are at least a century old) plausibly contribute towards their pre-eminence in research impact.

Contrasting regional performance is also observed in the third aspect of the World University Research Rankings: research collaborative-ness, with European universities emerging top in both industry collaboration and international collaboration. The dense network of European universities and strong academia-industry linkages [Horizon 2020 has a target of 20% for contribution allocated to SMEs under signed grants in Leadership in Enabling and Industrial Technologies (LEIT) and Societal Challenges[17]], underpinned by a long shared history of academic traditions, a modern framework of pan-European institutions e.g. European Research Council and well-funded research programmes such as Horizon 2020, plausibly contribute towards European universities' high % of international and industry collaboration.

3.3.2. Drill-down analysis into performance across seven individual indicators by location

To obtain higher resolution on the relative strengths and weaknesses between countries, further analysis is drilled down to the 7 individual indicators as shown in Tables 8, 9 and 10 below.

Table 8: Normalised location averages for research multi-disciplinarity indicators i.e. % multi-field and % multi-category indicators (figures in brackets denote number of universities represented in this study)

Location	Multi-field Norm. score	Multi-field % var. vs global	Multi-category Norm. score	Multi-category % var. vs global
USA (53)	46.6%	-9.6%	42.5%	-7.3%
UK (32)	43.6%	-15.4%	46.3%	1.0%
China (16)	70.7%	37.2%	46.1%	0.5%
Australia	53.0%	2.8%	51.3%	11.7%
Germany	45.9%	-10.9%	42.3%	-7.8%
Netherland	53.5%	3.8%	49.2%	7.3%
Canada	45.7%	-11.2%	47.2%	2.8%
Japan (10)	52.5%	2.0%	38.4%	-16.2%
France (8)	57.5%	11.7%	49.4%	7.6%
Korea (8)	67.5%	31.1%	43.0%	-6.4%
Switzerland	45.5%	-11.7%	43.7%	-4.8%
Italy (6)	56.1%	9.0%	50.3%	9.7%
Belgium	50.8%	-1.3%	47.2%	2.9%
Spain (5)	46.0%	-10.7%	45.2%	-1.4%
Sweden (5)	54.5%	5.7%	49.2%	7.3%
Malaysia	69.5%	34.9%	59.2%	29.0%
New	50.0%	-2.8%	48.5%	5.7%
Denmark	56.4%	9.4%	51.7%	12.6%
Russia (3)	42.6%	-17.3%	32.6%	-29.0%
India (3)	70.1%	36.1%	58.2%	26.8%
Singapore	74.1%	43.9%	52.0%	13.2%
Chile (2)	40.2%	-22.0%	38.7%	-15.7%
Brazil (2)	54.3%	5.5%	41.1%	-10.3%
Finland (2)	58.6%	13.8%	48.6%	5.9%
Norway (2)	34.1%	-33.9%	48.3%	5.3%
Israel (2)	44.6%	-13.5%	44.0%	-4.1%
Saudi	63.3%	22.8%	52.7%	14.9%
Ireland (2)	46.4%	-10.0%	41.0%	-10.6%
Mexico (2)	62.8%	22.0%	45.6%	-0.6%
Austria (2)	62.4%	21.1%	47.2%	2.9%
South	39.8%	-22.7%	50.0%	9.0%
Argentina	45.0%	-12.7%	40.3%	-12.1%
Thailand	59.4%	15.3%	48.3%	5.2%
Lebanon	42.8%	-16.9%	44.1%	-3.8%
Kazakhstan	44.6%	-13.3%	60.1%	31.0%
Colombia	44.0%	-14.6%	53.6%	16.8%
Global	51.5%	0.0%	45.9%	0.0%

Examining top performing countries in multi-field research, Singapore, Chinese, Indian and Korean universities have pushed significantly ahead, predominantly across the fields of "Natural sciences" and "Engineering and Technology" as they invest heavily in R&D to drive the development of their industries and economies. The Malaysian universities have the highest % multi-category research, predominantly in the field of "Engineering and Technology", especially in the category of "Materials engineering" (about a quarter of their multi-category publications).

Table 9: Normalised location averages for research impact indicators i.e. CNCI, % documents in top 1% and % documents in Q1 journals (figures in brackets denote number of universities represented in this study)

Location	CNCI Norm. score	CNCI % var. vs global	% docs in top 1% Norm. score	% docs in top 1% % var. vs global	% docs in Q1 journals Norm. score	% docs in Q1 journals % var. vs global
USA (53)	72.3%	22.5%	51.9%	36.0%	86.4%	14.3%
UK (32)	62.9%	6.7%	41.3%	8.3%	80.1%	5.9%
China (16)	51.4%	-12.9%	32.9%	-13.8%	73.1%	-3.3%
Australia	56.4%	-4.4%	34.6%	-9.3%	70.7%	-6.5%
Germany	60.9%	3.3%	39.5%	3.7%	75.8%	0.3%
Netherlan	64.0%	8.6%	42.8%	12.2%	83.0%	9.8%
Canada	60.1%	1.9%	36.7%	-3.8%	75.8%	0.2%
Japan (10)	40.1%	-32.0%	20.5%	-46.3%	65.8%	-13.0%
France (8)	55.8%	-5.4%	33.8%	-11.3%	74.3%	-1.8%
Korea (8)	47.3%	-19.8%	24.6%	-35.4%	66.6%	-11.9%
Switzerlan	68.3%	15.7%	48.2%	26.2%	82.3%	8.9%
Italy (6)	59.1%	0.2%	35.2%	-7.7%	77.9%	3.1%
Belgium	63.1%	7.0%	40.0%	4.9%	76.8%	1.5%
Spain (5)	56.5%	-4.1%	32.7%	-14.2%	75.5%	-0.2%
Sweden	60.3%	2.2%	37.1%	-2.7%	79.8%	5.6%
Malaysia	35.1%	-40.6%	15.8%	-58.6%	43.2%	-42.8%
New	52.2%	-11.6%	28.4%	-25.6%	66.0%	-12.7%
Denmark	64.5%	9.4%	41.3%	8.4%	81.3%	7.6%
Russia (3)	34.3%	-41.8%	17.1%	-55.2%	43.2%	-42.8%
India (3)	36.1%	-38.8%	13.9%	-63.7%	61.9%	-18.1%
Singapore	62.2%	5.5%	47.1%	23.5%	80.4%	6.4%
Chile (2)	43.2%	-26.8%	24.3%	-36.4%	61.6%	-18.6%
Brazil (2)	34.6%	-41.4%	16.3%	-57.2%	54.0%	-28.6%
Finland	60.6%	2.7%	39.7%	4.0%	78.1%	3.3%
Norway	62.4%	5.8%	41.0%	7.4%	74.6%	-1.3%
Israel (2)	53.5%	-9.3%	30.9%	-19.1%	73.5%	-2.8%
Saudi	51.8%	-12.2%	47.8%	25.2%	54.0%	-28.6%
Ireland (2)	55.2%	-6.4%	34.1%	-10.5%	73.8%	-2.4%
Mexico	32.4%	-45.0%	15.7%	-59.0%	52.5%	-30.6%
Austria	51.8%	-12.3%	30.4%	-20.2%	72.5%	-4.1%
South	61.4%	4.1%	43.0%	12.7%	66.4%	-12.2%
Argentina	36.3%	-38.5%	18.5%	-51.6%	61.2%	-19.0%
Thailand	36.1%	-38.8%	19.2%	-49.7%	53.3%	-29.5%
Lebanon	46.4%	-21.4%	28.5%	-25.3%	61.4%	-18.8%
Kazakhsta	19.5%	-66.9%	5.5%	-85.6%	22.7%	-70.0%
Colombia	50.8%	-13.9%	36.1%	-5.2%	68.6%	-9.2%
Global	59.0%	0.0%	38.1%	0.0%	75.6%	0.0%

Assessing research impact by location, US is clearly the *crème* de *la crème*, topping all three indicators of CNCI, % documents in top 1% and % documents in Q1 journals, followed by the Swiss, Dutch and Danish universities. American universities' open culture of welcoming diverse talent and renowned track record have helped them build a deep reservoir of talent that plausibly contributes towards their leading performance in these research impact indicators.

Table 10: Normalised location averages for research collaborative-ness indicators i.e.
% industry collaboration and % international collaboration
(figures in brackets denote number of universities represented in this study)

Location	Industry Collaboration	Industry Collaboration	International Collaboration	International Collaboration
USA (53)	30.4%	12.3%	39.9%	-22.8%
UK (32)	24.8%	-8.4%	56.4%	9.0%
China (16)	20.4%	-24.5%	36.7%	-29.0%
Australia (15)	14.1%	-48.1%	56.5%	9.3%
Germany (14)	36.1%	33.4%	58.1%	12.4%
Netherlands (12)	37.3%	37.9%	63.0%	21.8%
Canada (10)	24.9%	-7.8%	54.9%	6.2%
Japan (10)	34.0%	25.8%	34.2%	-33.8%
France (8)	47.1%	74.1%	59.9%	15.9%
Korea (8)	40.9%	51.3%	32.7%	-36.7%
Switzerland (7)	31.7%	17.0%	74.2%	43.4%
Italy (6)	23.1%	-14.6%	57.1%	10.4%
Belgium (5)	32.3%	19.4%	69.3%	34.0%
Spain (5)	21.9%	-19.0%	54.9%	6.2%
Sweden (5)	44.9%	66.1%	70.2%	35.6%
Malaysia (5)	4.3%	-84.2%	46.8%	-9.5%
New Zealand (4)	11.2%	-58.6%	64.2%	24.1%
Denmark (3)	47.5%	75.4%	68.2%	31.7%
Russia (3)	5.0%	-81.6%	47.4%	-8.3%
India (3)	15.6%	-42.2%	30.6%	-40.9%
Singapore (2)	27.4%	1.2%	67.1%	29.8%
Chile (2)	9.9%	-63.3%	65.1%	25.8%
Brazil (2)	9.9%	-63.4%	38.7%	-25.1%
Finland (2)	28.5%	5.2%	68.1%	31.6%
Norway (2)	26.5%	-2.0%	70.5%	36.3%
Israel (2)	19.7%	-27.2%	53.6%	3.7%
Saudi Arabia (2)	8.9%	-67.2%	80.5%	55.6%
Ireland (2)	20.1%	-25.7%	57.6%	11.4%
Mexico (2)	8.3%	-69.2%	53.1%	2.7%
Austria (2)	22.7%	-16.1%	70.5%	36.3%
South Africa (1)	19.2%	-28.9%	70.0%	35.2%
Argentina (1)	16.8%	-37.9%	50.3%	-2.7%
Thailand (1)	18.2%	-32.8%	50.5%	-2.4%
Lebanon (1)	22.9%	-15.4%	64.0%	23.6%
Kazakhstan (1)	1.6%	-94.2%	51.7%	-0.1%
Colombia (1)	6.5%	-76.1%	73.2%	41.5%
Global (250)	27.1%	0.0%	51.7%	0.0%

Evaluating countries in terms of industry collaboration, Danish, French, Swedish and Korean universities prominently stand out heads and shoulders above others, supported by their vibrant innovation-driven industries and strong academia-industry linkages. Swiss, Swedish, Belgian and Danish universities lead the pack in % international collaboration, all from countries predisposed to an outward-looking, international mindset from the impetus to transcend the small size of domestic markets.

3.4. *Correlation analysis on the indicators within each aspect as well as between the aspects*

Table 11: Correlation between indicators and aspects

Correlation between	Correlation coeff.
% multi-field documents and % multi-category documents	0.3232
CNCI and % documents in top 1%	0.9545
CNCI and % documents in Q1 journals	0.8538
% documents in top 1% and % documents in Q1 journals	0.8008
% industry collaboration and % international collaboration	0.0195
Research multi-disciplinary score and research impact score	-0.3774
Research multi-disciplinary score and research collaborative-ness score	0.0152
Research collaborative-ness score and research impact score	0.3701

4. Discussion

4.1. *Overall*

There is greater diversity among the top of the rankings with smaller, younger and less renowned universities from more countries making it to the top 10 (of which six are of small/medium size, as compared to two for QS 2020 rankings) and top 50 (of which sixteen are of small/medium size, as compared to eight for QS 2020 rankings). Six countries (USA, Netherlands, Denmark, France, Switzerland and Singapore) are represented in the top 10 as compared to 3 in QS (USA, UK, Switzerland). Dutch and Danish universities perform well across the board and bear further study to identify and understand the factors underpinning their all-round success. It is noteworthy that eight out of twelve Dutch universities are ranked in the top 50, the most from any location outside of the United States (16 out of 53) and double that of the United Kingdom (4 out of 32).

4.2. *Research multi-disciplinarity*

Asian universities have been found to be more multi-disciplinary than those from other regions, exemplified by Singapore, Chinese, Indian and Korean universities spearheading multi-field research, primarily across "Natural sciences" and "Engineering and Technology". With the advent of Industry 4.0 driving closer integration of multiple technologies straddling the aforementioned fields, these countries have dedicated significant resources to R&D in science and technology fields to enhance their competitiveness in moving up the global value chain and positioning themselves for the next wave of growth in the digital economy.

American and British universities, with 53 and 32 universities respectively represented in this study (making up about 34% of the study sample), performed below global averages in terms of research multi-disciplinarity. Traditions of specialization by field and organization by faculties/schools may be holding back American and European universities from fostering formation of multi-disciplinary teams to drive faster growth in multi-field research. This presents an interesting untapped opportunity which the universities could unlock to effect improvement in their overall multi-disciplinarity efforts.

Levitt[18] previously found that "the major difference between mono-disciplinarity and multi-disciplinarity indicates higher citation for mono-disciplinary articles". Noorden[19] also found that "Over three years, papers with diverse references tend to pick up fewer citations than the norm, but over 13 years they gain more. Some studies suggest that a little interdisciplinarity is better than a lot: papers that combine very disparate fields tend to earn fewer citations. But interdisciplinary work can have broad societal and economic impacts that are not captured by citations." A Yegros-Yegros[20] opined that "while combining multiple fields has a positive effect in knowledge creation, successful research is better achieved through research efforts that draw on a relatively proximal range of fields, as distal interdisciplinary research might be too risky and more likely to fail. On the other hand, these results may be interpreted as suggesting that scientific audiences are reluctant to cite heterodox papers that mix highly disparate bodies of knowledge—thus giving less credit to publications that are too groundbreaking or challenging".

4.3. *Research impact*

Relative to other countries, Swiss and American universities have high CNCI, % documents in top 1% and % documents in Q1 journals. The foremost driver is talent and the US has created a virtuous cycle with a conducive environment for world-class research - top talent supported with ample resources, first-rate infrastructure and robust intellectual property protection which generate best-in-class results that further boost the prestige and appeal of the US as a premier destination to draw in global talent who want to work with the best. Soh[21] has highlighted the influence of the English language on citations and by extension research impact. Further study is recommended to identify and understand the factors underpinning their research excellence.

Analysis of research impact performance by university size also revealed that small universities outperformed their bigger peers. Of the ten small universities, six (60%) of them were placed in the top 50 (top 20%). This bears testament to the axiom of quality rather than quantity, and that smaller universities are also able to punch above their weight in creating impact in the international research arena if they play according to their strengths.

4.4. *Research collaborative-ness*

European universities have been shown to be the most collaborative, leading the globe in forging industry and international partnerships for their research endeavours. Decades of European cooperation and integration under pan-European institutions such as the European Commission, European Research Council etc. have nurtured strong ties, long-standing partnerships and close collaboration among European universities and industries concerted efforts to drive academia-industry partnerships such as Horizon 2020' LEIT have also helped to contribute towards European universities' high % industry collaboration. National efforts to cultivate a strong culture of industry-academia collaboration through policy instruments also play a role, as exemplified by Danish, French and Swedish universities. According to an OECD report on University-Industry collaboration[22], OECD countries have implemented a variety of financial, regulatory and "soft" instruments, including but not limited to tax incentives, mobility schemes for researchers, networking events etc. to boost knowledge exchange between science and industry. It would be useful to delve further to identify and understand the factors underpinning strong academia-industry linkages.

Universities from smaller countries such as Switzerland, Sweden, Belgium and Denmark tend to collaborate more with international partners whereas those from bigger countries from India, Japan, China and US tend to collaborate less with international partners. This may be due to smaller countries lacking critical mass, therefore encouraging universities from these countries towards expanding their network of partnerships abroad.

References

1. QS methodology, https://www.topuniversities.com/qs-world-university-rankings/methodology (accessed on 10 Dec 2019)
2. THE methodology, https://www.timeshighereducation.com/world-university-rankings/world-university-rankings-2020-methodology (accessed on 10 Dec 2019)
3. ARWU methodology, http://www.shanghairanking.com/ARWU-Methodology-2019.html# (accessed on 10 Dec 2019)
4. Bahram Bekhradnia, 2016, "International university rankings: For good or ill?"
5. Philip G. Altbach, 2010, "The State of the Rankings"
6. Bowman et al., 2011, Anchoring effects in world university rankings: exploring biases in reputation scores, High Educ (2011) 61:431–444, DOI 10.1007/s10734-010-9339-1, http://www-personal.umich.edu/~bastedo/papers/BowmanBastedo.HE2011.pdf
7. Marope et al., 2013, Rankings and accountability in higher education: uses and misuses, https://unesdoc.unesco.org/ark:/48223/pf0000220789
8. K.S. Yeo et al., 2008, Integrated circuit design research ranking for worldwide universities, Journal of Circuits, Systems, and Computers Vol. 17, No. 1 (2008) 141–167
9. UNESCO science report, 2015, https://en.unesco.org/unesco_science_report
10. Alexander Riegler, Inclusive worldviews: Interdisciplinary research from a radical constructivist perspective, Worldviews, Science and Us, pp. 20-37 (2005), World Scientific
11. InCites Indicators Handbook, http://help.incites.clarivate.com/inCites2Live/8980-TRS/version/default/part/AttachmentData/data/InCites-Indicators-Handbook%20-%20June%202018.pdf
12. QS University classification, http://www.iu.qs.com/university-rankings/qs-classifications/, (accessed on 10 Dec 2019)
13. QS World University Rankings® 2020 report, https://www.topuniversities.com/student-info/qs-guides/qs-world-university-rankings-2020-supplement, (accessed on 10 Dec 2019)
14. Scuola Normale Superiore website, https://www.sns.it/en/scuola-normale-superiore/history, (accessed on 11 Dec 2019)
15. Sorbonne website, http://sorbonne-universite.fr/en/university/history-and-heritage, (accessed on 11 Dec 2019)
16. Telecom ParisTech website, https://internationaltelecomparistech.wp.imt.fr/incoming-students/few-words-about-telecom-paristech/, (accessed on 11 Dec 2019)
17. European Commission Horizon 2020 website, https://ec.europa.eu/programmes/horizon2020/sites/horizon2020/files/h2020_threeyearson_a4_horizontal_2018_web.pdf, accessed on 8 Jan 2020
18. Levitt et al., 2008, Is multidisciplinary research more highly cited? A macrolevel study, DOI: 10.1002/asi.20914
19. Noorden, 2015, Interdisciplinary research by the numbers, https://www.nature.com/news/interdisciplinary-research-by-the-numbers-1.18349

20. A Yegros-Yegros et al, 2015, Does Interdisciplinary Research Lead to Higher Citation Impact? The Different Effect of Proximal and Distal Interdisciplinarity, doi: 10.1371/journal.pone.0135095
21. Soh Kay Cheng, World university rankings: statistical issues and possible remedies, World Scientific
22. OECD, 2019, University-Industry Collaboration, New Evidence and Policy Options, https://doi.org/10.1787/e9c1e648-en

Appendices

Table 12: Global and regional rankings based on research multi-disciplinarity score

University	Location	Region	Multidiscipl. score	Multidiscipl. global rank	Multidiscipl. regional rank
Polytechnic University of Milan	Italy	Europe	100.0%	1	1
Universiti Teknologi Malaysia	Malaysia	Asia	99.5%	2	1
Technical University of Denmark	Denmark	Europe	98.8%	3	2
Hong Kong Polytechnic University	Hong Kong	Asia	98.0%	4	2
Delft University of Technology	Netherlands	Europe	94.6%	5	3
City University of Hong Kong	Hong Kong	Asia	93.9%	6	3
Royal Melbourne Institute of Technology (RMIT)	Australia	Oceania	93.6%	7	1
Loughborough University	UK	Europe	91.4%	8	4
University of Twente	Netherlands	Europe	90.8%	9	5
Indian Institute of Technology (IIT) - Delhi	India	Asia	90.7%	10	4
Scuola Superiore Sant'Anna	Italy	Europe	90.7%	11	6
Nanyang Technological University	Singapore	Asia	90.2%	12	5
National Chiao Tung University	Taiwan	Asia	90.1%	13	6
Korea Advanced Institute of Science & Technology (KAIST)	Korea	Asia	89.5%	14	7
Eindhoven University of Technology	Netherlands	Europe	89.4%	15	7
Universiti Putra Malaysia	Malaysia	Asia	87.6%	16	8
Georgia Institute of Technology	USA	N America	87.6%	17	1
Universiti Sains Malaysia	Malaysia	Asia	87.5%	18	9
National Tsing Hua University	Taiwan	Asia	87.0%	19	10
Tecnologico de Monterrey	Mexico	N America	86.4%	20	2
Ecole des Ponts ParisTech	France	Europe	86.4%	21	8

Table 12: (*Continued*)

University	Location	Region	Multidiscipl. score	Multidiscipl. global rank	Multidiscipl. regional rank
CentraleSupelec	France	Europe	86.0%	22	9
National Cheng Kung University	Taiwan	Asia	85.6%	23	11
Tsinghua University	China (Mainland)	Asia	85.6%	24	12
Chalmers University of Technology	Sweden	Europe	85.4%	25	10
University of Technology Sydney	Australia	Oceania	84.8%	26	2
King Fahd University of Petroleum & Minerals	Saudi Arabia	Asia	84.7%	27	13
Technical University of Berlin	Germany	Europe	84.6%	28	11
Aalto University	Finland	Europe	84.2%	29	12
Vienna University of Technology	Austria	Europe	84.0%	30	13
Telecom ParisTech	France	Europe	83.9%	31	14
Royal Institute of Technology	Sweden	Europe	83.9%	32	15
University of Wollongong	Australia	Oceania	83.7%	33	3
Pohang University of Science & Technology (POSTECH)	Korea	Asia	83.5%	34	14
Ecole Polytechnique Federale de Lausanne	Switzerland	Europe	83.3%	35	16
Indian Institute of Technology (IIT) - Bombay	India	Asia	83.1%	36	15
Hong Kong University of Science & Technology	Hong Kong	Asia	82.2%	37	16
University of Waterloo	Canada	N America	82.1%	38	3
Institut d'Etudes Politiques Paris (SciencePo)	France	Europe	81.5%	39	17
Wageningen University & Research	Netherlands	Europe	81.0%	40	18
Hanyang University	Korea	Asia	79.9%	41	17
Arizona State University	USA	N America	78.7%	42	4
Indian Institute of Science (IISC) - Bangalore	India	Asia	77.8%	43	18

Table 12: (*Continued*)

University	Location	Region	Multidiscipl. score	Multidiscipl. global rank	Multidiscipl. regional rank
London School Economics & Political Science	UK	Europe	77.5%	44	19
Karlsruhe Institute of Technology	Germany	Europe	77.0%	45	20
University of Bath	UK	Europe	76.4%	46	21
Universiti Kebangsaan Malaysia	Malaysia	Asia	76.2%	47	19
Zhejiang University	China (Mainland)	Asia	76.1%	48	20
Purdue University	USA	N America	75.6%	49	5
Carnegie Mellon University	USA	N America	75.5%	50	6
National University of Singapore	Singapore	Asia	74.7%	51	21
Shanghai Jiao Tong University	China (Mainland)	Asia	73.7%	52	22
Curtin University	Australia	Oceania	73.2%	53	4
University of Science & Technology of China	China (Mainland)	Asia	73.0%	54	23
ETH Zurich	Switzerland	Europe	72.1%	55	22
Queensland University of Technology (QUT)	Australia	Oceania	71.8%	56	5
Korea University	Korea	Asia	70.8%	57	24
Chulalongkorn University	Thailand	Asia	70.4%	58	25
University of Illinois Urbana-Champaign	USA	N America	70.3%	59	7
University of New South Wales Sydney	Australia	Oceania	70.2%	60	6
University of Maryland College Park	USA	N America	70.1%	61	8
Waseda University	Japan	Asia	69.9%	62	26
Universiti Malaya	Malaysia	Asia	69.8%	63	27

Table 12: (*Continued*)

University	Location	Region	Multidiscipl. score	Multidiscipl. global rank	Multidiscipl. regional rank
RWTH Aachen University	Germany	Europe	68.7%	64	23
National Taiwan University	Taiwan	Asia	68.7%	65	28
Texas A&M University College Station	USA	N America	68.5%	66	9
Al-Farabi Kazakh National University	Kazakhstan	Asia	68.5%	67	29
Rice University	USA	N America	68.3%	68	10
Victoria University Wellington	New Zealand	Oceania	68.2%	69	7
University of Texas Austin	USA	N America	68.0%	70	11
Kyung Hee University	Korea	Asia	68.0%	71	30
Tokyo Institute of Technology	Japan	Asia	67.8%	72	31
University of Canterbury	New Zealand	Oceania	67.2%	73	8
King Abdulaziz University	Saudi Arabia	Asia	67.0%	74	32
Universidade Estadual de Campinas	Brazil	S America	66.9%	75	1
Technical University of Munich	Germany	Europe	66.7%	76	24
Ghent University	Belgium	Europe	66.7%	77	25
Vrije Universiteit Brussel	Belgium	Europe	66.5%	78	26
University of Antwerp	Belgium	Europe	66.4%	79	27
University of Auckland	New Zealand	Oceania	66.1%	80	9
Lancaster University	UK	Europe	65.8%	81	28
University of Hong Kong	Hong Kong	Asia	65.6%	82	33
Penn State University	USA	N America	65.5%	83	12
Tohoku University	Japan	Asia	65.1%	84	34
University of Nottingham	UK	Europe	65.0%	85	29
Michigan State University	USA	N America	65.0%	86	13

Table 12: (*Continued*)

University	Location	Region	Multidiscipl. score	Multidiscipl. global rank	Multidiscipl. regional rank
Dresden University of Technology	Germany	Europe	64.8%	87	30
University of Queensland	Australia	Oceania	64.8%	88	10
KU Leuven	Belgium	Europe	64.5%	89	31
University of Newcastle	Australia	Oceania	64.2%	90	11
University of Reading	UK	Europe	64.2%	91	32
Sungkyunkwan University (SKKU)	Korea	Asia	64.1%	92	35
University of California Santa Barbara	USA	N America	63.8%	93	14
University of the Andes Colombia	Colombia	S America	63.8%	94	2
Autonomous University of Barcelona	Spain	Europe	63.6%	95	33
Massachusetts Institute of Technology (MIT)	USA	N America	63.4%	96	15
University of Navarra	Spain	Europe	63.3%	97	34
Monash University	Australia	Oceania	62.5%	98	12
University of Southampton	UK	Europe	62.5%	99	35
Seoul National University (SNU)	Korea	Asia	62.3%	100	36
University of Colorado Boulder	USA	N America	62.2%	101	16
University of Bologna	Italy	Europe	62.2%	102	36
Nanjing University	China (Mainland)	Asia	62.1%	103	37
McGill University	Canada	N America	61.9%	104	17
Queens University - Canada	Canada	N America	61.5%	105	18
Vrije Universiteit Amsterdam	Netherlands	Europe	61.4%	106	37
Complutense University of Madrid	Spain	Europe	61.4%	107	38
University of California Berkeley	USA	N America	61.3%	108	19

Table 12: (*Continued*)

University	Location	Region	Multidiscipl. score	Multidiscipl. global rank	Multidiscipl. regional rank
University of Adelaide	Australia	Oceania	61.2%	109	13
Macquarie University	Australia	Oceania	61.1%	110	14
Peking University	China (Mainland)	Asia	60.9%	111	38
University of Sheffield	UK	Europe	60.7%	112	39
Chinese University of Hong Kong	Hong Kong	Asia	60.6%	113	39
Kyushu University	Japan	Asia	60.6%	114	40
University of York - UK	UK	Europe	60.5%	115	40
Newcastle University - UK	UK	Europe	60.4%	116	41
Queens University Belfast	UK	Europe	60.4%	117	42
Utrecht University	Netherlands	Europe	60.0%	118	43
University of Florida	USA	N America	60.0%	119	20
Trinity College Dublin	Ireland	Europe	59.9%	120	44
University of British Columbia	Canada	N America	59.7%	121	21
University of Southern California	USA	N America	59.7%	122	22
University of Minnesota Twin Cities	USA	N America	59.7%	123	23
Yonsei University	Korea	Asia	59.7%	124	41
Radboud University Nijmegen	Netherlands	Europe	59.6%	125	45
Sapienza University Rome	Italy	Europe	59.4%	126	46
Aarhus University	Denmark	Europe	59.4%	127	47
University of Notre Dame	USA	N America	59.3%	128	24
University of Vienna	Austria	Europe	59.3%	129	48
Western University (University of Western Ontario)	Canada	N America	59.2%	130	25

Table 12: (*Continued*)

University	Location	Region	Multidiscipl. score	Multidiscipl. global rank	Multidiscipl. regional rank
University of Alberta	Canada	N America	59.2%	131	26
Australian National University	Australia	Oceania	59.0%	132	15
Hebrew University of Jerusalem	Israel	Asia	59.0%	133	42
University of Melbourne	Australia	Oceania	58.8%	134	16
University of Illinois Chicago	USA	N America	58.8%	135	27
University of Cape Town	South Africa	Africa	58.7%	136	1
Fudan University	China (Mainland)	Asia	58.7%	137	43
Ecole Polytechnique	France	Europe	58.7%	138	49
University of Wisconsin Madison	USA	N America	58.7%	139	28
University of Montreal	Canada	N America	58.6%	140	29
University of Sussex	UK	Europe	58.4%	141	50
University of California Davis	USA	N America	58.1%	142	30
University of Calgary	Canada	N America	58.0%	143	31
Universidade de Sao Paulo	Brazil	S America	57.9%	144	3
University of Warwick	UK	Europe	57.9%	145	51
University of Padua	Italy	Europe	57.8%	146	52
Cardiff University	UK	Europe	57.8%	147	53
Lund University	Sweden	Europe	57.3%	148	54
University of Aberdeen	UK	Europe	57.3%	149	55
University of Exeter	UK	Europe	57.0%	150	56
Brown University	USA	N America	57.0%	151	32
Imperial College London	UK	Europe	57.0%	152	57

Table 12: (*Continued*)

University	Location	Region	Multidiscipl. score	Multidiscipl. global rank	Multidiscipl. regional rank
Kings College London	UK	Europe	56.9%	153	58
Tel Aviv University	Israel	Asia	56.8%	154	44
American University of Beirut	Lebanon	Asia	56.8%	155	45
University College London	UK	Europe	56.8%	156	59
University of Pittsburgh	USA	N America	56.7%	157	33
University of Western Australia	Australia	Oceania	56.7%	158	17
Maastricht University	Netherlands	Europe	56.7%	159	60
Stockholm University	Sweden	Europe	56.6%	160	61
McMaster University	Canada	N America	56.6%	161	34
Ohio State University	USA	N America	56.5%	162	35
University of Groningen	Netherlands	Europe	56.5%	163	62
University of Bristol	UK	Europe	56.4%	164	63
University of Sydney	Australia	Oceania	56.4%	165	18
Universite Catholique Louvain	Belgium	Europe	56.4%	166	64
Hokkaido University	Japan	Asia	56.3%	167	46
University of North Carolina Chapel Hill	USA	N America	56.2%	168	36
Osaka University	Japan	Asia	56.2%	169	47
University of Otago	New Zealand	Oceania	56.2%	170	19
University of Michigan	USA	N America	56.0%	171	37
University of California Irvine	USA	N America	56.0%	172	38
University of Helsinki	Finland	Europe	56.0%	173	65
University of Gottingen	Germany	Europe	55.8%	174	66
University of Buenos Aires	Argentina	S America	55.8%	175	4

Table 12: (*Continued*)

University	Location	Region	Multidiscipl. score	Multidiscipl. global rank	Multidiscipl. regional rank
Uppsala University	Sweden	Europe	55.7%	176	67
University of Leeds	UK	Europe	55.6%	177	68
University of Manchester	UK	Europe	55.5%	178	69
PSL Research University Paris (ComUE)	France	Europe	55.5%	179	70
Ecole Normale Superieure de Lyon (ENS LYON)	France	Europe	55.5%	180	71
University of Tokyo	Japan	Asia	55.4%	181	48
University of Cambridge	UK	Europe	55.4%	182	72
Universidad Nacional Autonoma de Mexico	Mexico	N America	55.4%	183	39
University of California San Diego	USA	N America	55.3%	184	40
Kyoto University	Japan	Asia	55.3%	185	49
Northwestern University	USA	N America	55.2%	186	41
University of Basel	Switzerland	Europe	55.2%	187	73
University of Barcelona	Spain	Europe	55.2%	188	74
Saint Petersburg State University	Russia	Europe	55.2%	189	75
Stanford University	USA	N America	54.9%	190	42
Keio University	Japan	Asia	54.9%	191	50
Autonomous University of Madrid	Spain	Europe	54.8%	192	76
Universidad de Chile	Chile	S America	54.6%	193	5
Boston University	USA	N America	54.6%	194	43
University College Dublin	Ireland	Europe	54.3%	195	77
University of Bergen	Norway	Europe	53.9%	196	78
University of Amsterdam	Netherlands	Europe	53.9%	197	79
University of Oslo	Norway	Europe	53.7%	198	80

Table 12: (*Continued*)

University	Location	Region	Multidiscipl. score	Multidiscipl. global rank	Multidiscipl. regional rank
University of Copenhagen	Denmark	Europe	53.7%	199	81
Columbia University	USA	N America	53.5%	200	44
University of Washington Seattle	USA	N America	53.5%	201	45
Georgetown University	USA	N America	53.5%	202	46
Vanderbilt University	USA	N America	53.5%	203	47
Nagoya University	Japan	Asia	53.4%	204	51
University of Zurich	Switzerland	Europe	53.2%	205	82
Cornell University	USA	N America	53.0%	206	48
Erasmus University Rotterdam	Netherlands	Europe	52.8%	207	83
Free University of Berlin	Germany	Europe	52.8%	208	84
Case Western Reserve University	USA	N America	52.6%	209	49
University of California Los Angeles	USA	N America	52.5%	210	50
Princeton University	USA	N America	52.4%	211	51
University of Pennsylvania	USA	N America	52.2%	212	52
University of Virginia	USA	N America	52.2%	213	53
University of St Andrews	UK	Europe	52.1%	214	85
University of Birmingham	UK	Europe	51.8%	215	86
New York University	USA	N America	51.7%	216	54
Humboldt University of Berlin	Germany	Europe	51.6%	217	87
Sorbonne Universite	France	Europe	51.6%	218	88
University of Oxford	UK	Europe	51.6%	219	89
University of Edinburgh	UK	Europe	51.5%	220	90
Yale University	USA	N America	51.3%	221	55

Table 12: (*Continued*)

University	Location	Region	Multidiscipl. score	Multidiscipl. global rank	Multidiscipl. regional rank
Durham University	UK	Europe	51.3%	222	91
University of Liverpool	UK	Europe	51.2%	223	92
Duke University	USA	N America	51.1%	224	56
University of Freiburg	Germany	Europe	51.0%	225	93
Novosibirsk State University	Russia	Europe	51.0%	226	94
University of Toronto	Canada	N America	50.7%	227	57
University of Chicago	USA	N America	50.6%	228	58
University of Lausanne	Switzerland	Europe	50.3%	229	95
Johns Hopkins University	USA	N America	50.2%	230	59
Queen Mary University London	UK	Europe	50.0%	231	96
Dartmouth College	USA	N America	49.9%	232	60
Emory University	USA	N America	49.1%	233	61
University of Munich	Germany	Europe	49.0%	234	97
Leiden University	Netherlands	Europe	49.0%	235	98
University of Glasgow	UK	Europe	48.6%	236	99
Pontificia Universidad Catolica de Chile	Chile	S America	48.6%	237	6
University of Hamburg	Germany	Europe	48.4%	238	100
University of Geneva	Switzerland	Europe	48.1%	239	101
Eberhard Karls University of Tubingen	Germany	Europe	48.1%	240	102
California Institute of Technology	USA	N America	47.9%	241	62
Washington University (WUSTL)	USA	N America	47.6%	242	63
Scuola Normale Superiore di Pisa	Italy	Europe	47.6%	243	103
University of Rochester	USA	N America	46.7%	244	64

Table 12: (*Continued*)

University	Location	Region	Multidiscipl. score	Multidiscipl. global rank	Multidiscipl. regional rank
University of Bonn	Germany	Europe	46.5%	245	104
Harvard University	USA	N America	46.3%	246	65
University of Leicester	UK	Europe	45.9%	247	105
University of Bern	Switzerland	Europe	45.7%	248	106
Ruprecht Karls University Heidelberg	Germany	Europe	42.1%	249	107
Lomonosov Moscow State University	Russia	Europe	41.2%	250	108

Table 13: Global and regional rankings by research impact score

University	Location	Region	Impact score	Impact global rank	Impact regional rank
Massachusetts Institute of Technology (MIT)	USA	N America	100.0%	1	1
Stanford University	USA	N America	88.1%	2	2
Harvard University	USA	N America	87.6%	3	3
California Institute of Technology	USA	N America	87.1%	4	4
University of California Berkeley	USA	N America	83.7%	5	5
Princeton University	USA	N America	83.6%	6	6
Carnegie Mellon University	USA	N America	82.2%	7	7
University of Chicago	USA	N America	80.8%	8	8
Washington University (WUSTL)	USA	N America	80.7%	9	9
Yale University	USA	N America	78.0%	10	10
University of California Santa Barbara	USA	N America	77.9%	11	11
University of Washington Seattle	USA	N America	77.7%	12	12
Duke University	USA	N America	77.6%	13	13
University of California San Diego	USA	N America	77.2%	14	14
University of Pennsylvania	USA	N America	76.2%	15	15
Northwestern University	USA	N America	76.0%	16	16
Scuola Normale Superiore di Pisa	Italy	Europe	75.9%	17	1
Ecole Polytechnique Federale de Lausanne	Switzerland	Europe	75.5%	18	2
Columbia University	USA	N America	75.4%	19	17
Rice University	USA	N America	75.2%	20	18
Johns Hopkins University	USA	N America	74.8%	21	19

Table 13: (*Continued*)

University	Location	Region	Impact score	Impact global rank	Impact regional rank
Cornell University	USA	N America	74.1%	22	20
Vanderbilt University	USA	N America	73.9%	23	21
University of California Los Angeles	USA	N America	73.7%	24	22
University of Oxford	UK	Europe	73.5%	25	3
ETH Zurich	Switzerland	Europe	72.6%	26	4
Queen Mary University London	UK	Europe	72.4%	27	5
University of North Carolina Chapel Hill	USA	N America	72.2%	28	23
University of Cambridge	UK	Europe	72.1%	29	6
Erasmus University Rotterdam	Netherlands	Europe	71.8%	30	7
Boston University	USA	N America	71.7%	31	24
University of Colorado Boulder	USA	N America	71.5%	32	25
University of Edinburgh	UK	Europe	71.4%	33	8
Wageningen University & Research	Netherlands	Europe	71.3%	34	9
Hong Kong University of Science & Technology	Hong Kong	Asia	70.9%	35	1
Emory University	USA	N America	70.7%	36	26
Imperial College London	UK	Europe	70.5%	37	10
University of California Irvine	USA	N America	69.9%	38	27
New York University	USA	N America	69.7%	39	28
University of Michigan	USA	N America	69.5%	40	29
University of Maryland College Park	USA	N America	69.4%	41	30
University of Sussex	UK	Europe	68.8%	42	11
University of Southern California	USA	N America	68.3%	43	31
University College London	UK	Europe	68.2%	44	12

Table 13: (*Continued*)

University	Location	Region	Impact score	Impact global rank	Impact regional rank
University of Amsterdam	Netherlands	Europe	68.1%	45	13
University of Toronto	Canada	N America	67.8%	46	32
University of Rochester	USA	N America	67.8%	47	33
Stockholm University	Sweden	Europe	67.6%	48	14
University of Pittsburgh	USA	N America	67.6%	49	34
University of Texas Austin	USA	N America	67.3%	50	35
Radboud University Nijmegen	Netherlands	Europe	67.2%	51	15
Ruprecht Karls University Heidelberg	Germany	Europe	67.2%	52	16
Dartmouth College	USA	N America	67.2%	53	36
Georgia Institute of Technology	USA	N America	66.7%	54	37
KU Leuven	Belgium	Europe	66.7%	55	17
University of Zurich	Switzerland	Europe	66.6%	56	18
Leiden University	Netherlands	Europe	66.3%	57	19
Kings College London	UK	Europe	66.2%	58	20
Utrecht University	Netherlands	Europe	66.1%	59	21
Nanyang Technological University	Singapore	Asia	65.9%	60	2
University of Bristol	UK	Europe	65.9%	61	22
Chinese University of Hong Kong	Hong Kong	Asia	65.9%	62	3
Brown University	USA	N America	65.9%	63	38
University of Copenhagen	Denmark	Europe	65.7%	64	23
University of Geneva	Switzerland	Europe	65.6%	65	24
University of Groningen	Netherlands	Europe	65.6%	66	25
Vrije Universiteit Amsterdam	Netherlands	Europe	65.6%	67	26

Table 13: (*Continued*)

University	Location	Region	Impact score	Impact global rank	Impact regional rank
PSL Research University Paris (ComUE)	France	Europe	65.6%	68	27
Ohio State University	USA	N America	65.5%	69	39
University of Glasgow	UK	Europe	65.4%	70	28
King Abdulaziz University	Saudi Arabia	Asia	65.0%	71	4
University of Wisconsin Madison	USA	N America	64.8%	72	40
Universite Catholique Louvain	Belgium	Europe	64.6%	73	29
University of Illinois Urbana-Champaign	USA	N America	64.1%	74	41
Case Western Reserve University	USA	N America	64.0%	75	42
University of Virginia	USA	N America	63.9%	76	43
University of Munich	Germany	Europe	63.5%	77	30
University of Birmingham	UK	Europe	63.5%	78	31
University of Basel	Switzerland	Europe	63.4%	79	32
University of Minnesota Twin Cities	USA	N America	63.4%	80	44
University of Exeter	UK	Europe	63.3%	81	33
University of Bonn	Germany	Europe	63.2%	82	34
McMaster University	Canada	N America	63.1%	83	45
University of Barcelona	Spain	Europe	62.9%	84	35
University of California Davis	USA	N America	62.8%	85	46
Sorbonne Universite	France	Europe	62.6%	86	36
Georgetown University	USA	N America	62.6%	87	47
Technical University of Denmark	Denmark	Europe	62.6%	88	37
Lancaster University	UK	Europe	62.5%	89	38
McGill University	Canada	N America	62.5%	90	48

Table 13: (*Continued*)

University	Location	Region	Impact score	Impact global rank	Impact regional rank
University of Southampton	UK	Europe	62.5%	91	39
University of Lausanne	Switzerland	Europe	62.5%	92	40
University of Bern	Switzerland	Europe	62.1%	93	41
Cardiff University	UK	Europe	62.1%	94	42
University of British Columbia	Canada	N America	62.1%	95	49
University of Leicester	UK	Europe	61.9%	96	43
National University of Singapore	Singapore	Asia	61.8%	97	5
University of Liverpool	UK	Europe	61.8%	98	44
Uppsala University	Sweden	Europe	61.8%	99	45
Technical University of Munich	Germany	Europe	61.7%	100	46
Newcastle University - UK	UK	Europe	61.6%	101	47
Humboldt University of Berlin	Germany	Europe	61.4%	102	48
University of Manchester	UK	Europe	61.0%	103	49
University of Oslo	Norway	Europe	60.9%	104	50
Aarhus University	Denmark	Europe	60.7%	105	51
University of Freiburg	Germany	Europe	60.7%	106	52
University of Helsinki	Finland	Europe	60.5%	107	53
University of Adelaide	Australia	Oceania	60.3%	108	1
Arizona State University	USA	N America	60.3%	109	50
Lund University	Sweden	Europe	60.2%	110	54
Eberhard Karls University of Tubingen	Germany	Europe	59.9%	111	55
University of Warwick	UK	Europe	59.7%	112	56
Aalto University	Finland	Europe	59.6%	113	57

Table 13: (*Continued*)

University	Location	Region	Impact score	Impact global rank	Impact regional rank
University of Aberdeen	UK	Europe	59.5%	114	58
University of Melbourne	Australia	Oceania	59.4%	115	2
Durham University	UK	Europe	59.3%	116	59
City University of Hong Kong	Hong Kong	Asia	59.3%	117	6
Australian National University	Australia	Oceania	59.2%	118	3
Ecole Polytechnique	France	Europe	59.1%	119	60
University of Bologna	Italy	Europe	59.1%	120	61
University of Notre Dame	USA	N America	59.0%	121	51
University of Leeds	UK	Europe	59.0%	122	62
University of Hamburg	Germany	Europe	59.0%	123	63
University of Navarra	Spain	Europe	58.9%	124	64
University of Padua	Italy	Europe	58.9%	125	65
University of Bergen	Norway	Europe	58.9%	126	66
University of Science & Technology of China	China (Mainland)	Asia	58.8%	127	7
Penn State University	USA	N America	58.7%	128	52
University of Sydney	Australia	Oceania	58.5%	129	4
Maastricht University	Netherlands	Europe	58.3%	130	67
University of Antwerp	Belgium	Europe	58.1%	131	68
Free University of Berlin	Germany	Europe	58.0%	132	69
University of Gottingen	Germany	Europe	57.8%	133	70
University of Hong Kong	Hong Kong	Asia	57.7%	134	8
University of Sheffield	UK	Europe	57.7%	135	71

Table 13: (*Continued*)

University	Location	Region	Impact score	Impact global rank	Impact regional rank
Ecole Normale Superieure de Lyon (ENS LYON)	France	Europe	57.6%	136	72
University of York - UK	UK	Europe	57.6%	137	73
University of Queensland	Australia	Oceania	57.6%	138	5
University of Cape Town	South Africa	Africa	57.5%	139	1
University of St Andrews	UK	Europe	57.5%	140	74
Michigan State University	USA	N America	57.4%	141	53
Dresden University of Technology	Germany	Europe	57.2%	142	75
University of Nottingham	UK	Europe	57.2%	143	76
University of Calgary	Canada	N America	57.2%	144	54
University of Montreal	Canada	N America	57.2%	145	55
University of Technology Sydney	Australia	Oceania	57.0%	146	6
University of Reading	UK	Europe	56.8%	147	77
Autonomous University of Madrid	Spain	Europe	56.8%	148	78
Vrije Universiteit Brussel	Belgium	Europe	56.7%	149	79
Hong Kong Polytechnic University	Hong Kong	Asia	56.7%	150	9
Royal Institute of Technology	Sweden	Europe	56.6%	151	80
Ghent University	Belgium	Europe	56.6%	152	81
Tsinghua University	China (Mainland)	Asia	56.3%	153	10
Autonomous University of Barcelona	Spain	Europe	56.2%	154	82
University of Western Australia	Australia	Oceania	56.1%	155	7
University of Alberta	Canada	N America	56.0%	156	56
Eindhoven University of Technology	Netherlands	Europe	56.0%	157	83

Table 13: (*Continued*)

University	Location	Region	Impact score	Impact global rank	Impact regional rank
University of Twente	Netherlands	Europe	55.9%	158	84
Ecole des Ponts ParisTech	France	Europe	55.8%	159	85
Karlsruhe Institute of Technology	Germany	Europe	55.4%	160	86
University of New South Wales Sydney	Australia	Oceania	55.3%	161	8
University College Dublin	Ireland	Europe	55.2%	162	87
Pohang University of Science & Technology (POSTECH)	Korea	Asia	55.2%	163	11
Queens University Belfast	UK	Europe	55.0%	164	88
University of Florida	USA	N America	55.0%	165	57
Monash University	Australia	Oceania	54.9%	166	9
Delft University of Technology	Netherlands	Europe	54.8%	167	89
Purdue University	USA	N America	54.7%	168	58
London School Economics & Political Science	UK	Europe	54.7%	169	90
Trinity College Dublin	Ireland	Europe	54.7%	170	91
RWTH Aachen University	Germany	Europe	54.3%	171	92
Telecom ParisTech	France	Europe	53.7%	172	93
Hebrew University of Jerusalem	Israel	Asia	53.7%	173	12
Peking University	China (Mainland)	Asia	53.6%	174	13
University of Auckland	New Zealand	Oceania	53.5%	175	10
University of Vienna	Austria	Europe	53.1%	176	94
University of Bath	UK	Europe	53.0%	177	95
Scuola Superiore Sant'Anna	Italy	Europe	53.0%	178	96
University of Wollongong	Australia	Oceania	52.8%	179	11

Table 13: (*Continued*)

University	Location	Region	Impact score	Impact global rank	Impact regional rank
University of Twente	Netherlands	Europe	55.9%	158	84
Ecole des Ponts ParisTech	France	Europe	55.8%	159	85
Karlsruhe Institute of Technology	Germany	Europe	55.4%	160	86
University of New South Wales Sydney	Australia	Oceania	55.3%	161	8
University College Dublin	Ireland	Europe	55.2%	162	87
Pohang University of Science & Technology (POSTECH)	Korea	Asia	55.2%	163	11
Queens University Belfast	UK	Europe	55.0%	164	88
University of Florida	USA	N America	55.0%	165	57
Monash University	Australia	Oceania	54.9%	166	9
Delft University of Technology	Netherlands	Europe	54.8%	167	89
Purdue University	USA	N America	54.7%	168	58
London School Economics & Political Science	UK	Europe	54.7%	169	90
Trinity College Dublin	Ireland	Europe	54.7%	170	91
RWTH Aachen University	Germany	Europe	54.3%	171	92
Telecom ParisTech	France	Europe	53.7%	172	93
Hebrew University of Jerusalem	Israel	Asia	53.7%	173	12
Peking University	China (Mainland)	Asia	53.6%	174	13
University of Auckland	New Zealand	Oceania	53.5%	175	10
University of Vienna	Austria	Europe	53.1%	176	94
University of Bath	UK	Europe	53.0%	177	95
Scuola Superiore Sant'Anna	Italy	Europe	53.0%	178	96
University of Wollongong	Australia	Oceania	52.8%	179	11

Table 13: (*Continued*)

University	Location	Region	Impact score	Impact global rank	Impact regional rank
University of Waterloo	Canada	N America	52.7%	180	59
Tel Aviv University	Israel	Asia	52.6%	181	14
University of the Andes Colombia	Colombia	S America	52.4%	182	1
Korea Advanced Institute of Science & Technology (KAIST)	Korea	Asia	52.3%	183	15
Queens University - Canada	Canada	N America	52.1%	184	60
Chalmers University of Technology	Sweden	Europe	52.1%	185	97
University of Illinois Chicago	USA	N America	51.9%	186	61
Sapienza University Rome	Italy	Europe	51.9%	187	98
Texas A&M University College Station	USA	N America	51.7%	188	62
Sungkyunkwan University (SKKU)	Korea	Asia	51.7%	189	16
Technical University of Berlin	Germany	Europe	51.5%	190	99
Macquarie University	Australia	Oceania	51.4%	191	12
Vienna University of Technology	Austria	Europe	51.1%	192	100
National Tsing Hua University	Taiwan	Asia	50.8%	193	17
Nanjing University	China (Mainland)	Asia	50.5%	194	18
Western University (University of Western Ontario)	Canada	N America	50.2%	195	63
National Taiwan University	Taiwan	Asia	50.2%	196	19
Curtin University	Australia	Oceania	50.0%	197	13
Seoul National University (SNU)	Korea	Asia	50.0%	198	20
University of Canterbury	New Zealand	Oceania	50.0%	199	14
Queensland University of Technology (QUT)	Australia	Oceania	49.9%	200	15

Table 13: (*Continued*)

University	Location	Region	Impact score	Impact global rank	Impact regional rank
University of Otago	New Zealand	Oceania	49.7%	201	16
University of Tokyo	Japan	Asia	49.3%	202	21
Polytechnic University of Milan	Italy	Europe	49.1%	203	101
Fudan University	China (Mainland)	Asia	49.0%	204	22
CentraleSupelec	France	Europe	48.4%	205	102
Pontificia Universidad Catolica de Chile	Chile	S America	47.7%	206	2
University of Newcastle	Australia	Oceania	47.3%	207	17
Royal Melbourne Institute of Technology (RMIT)	Australia	Oceania	46.8%	208	18
Kyoto University	Japan	Asia	46.7%	209	23
American University of Beirut	Lebanon	Asia	45.9%	210	24
Shanghai Jiao Tong University	China (Mainland)	Asia	45.1%	211	25
Yonsei University	Korea	Asia	45.0%	212	26
Tokyo Institute of Technology	Japan	Asia	44.7%	213	27
Victoria University Wellington	New Zealand	Oceania	44.2%	214	19
Loughborough University	UK	Europe	44.0%	215	103
Zhejiang University	China (Mainland)	Asia	43.4%	216	28
Korea University	Korea	Asia	43.0%	217	29
Nagoya University	Japan	Asia	42.7%	218	30
Osaka University	Japan	Asia	42.7%	219	31
Complutense University of Madrid	Spain	Europe	42.6%	220	104

Table 13: (*Continued*)

University	Location	Region	Impact score	Impact global rank	Impact regional rank
Universidad Nacional Autonoma de Mexico	Mexico	N America	31.1%	244	65
Lomonosov Moscow State University	Russia	Europe	30.8%	245	107
Universiti Sains Malaysia	Malaysia	Asia	29.8%	246	48
Saint Petersburg State University	Russia	Europe	28.4%	247	108
Universiti Kebangsaan Malaysia	Malaysia	Asia	28.2%	248	49
Universiti Putra Malaysia	Malaysia	Asia	27.9%	249	50
Al-Farabi Kazakh National University	Kazakhstan	Asia	16.1%	250	51

Table 14: Global and regional rankings by research collaborative-ness score

University	Location	Region	Collaborative-ness score	Collaborat. global rank	Collaborat. regional rank
CentraleSupelec	France	Europe	100.0%	1	1
Telecom ParisTech	France	Europe	92.2%	2	2
Eindhoven University of Technology	Netherlands	Europe	89.6%	3	3
Chalmers University of Technology	Sweden	Europe	88.7%	4	4
Technical University of Denmark	Denmark	Europe	80.4%	5	5
University of Basel	Switzerland	Europe	78.9%	6	6
Royal Institute of Technology	Sweden	Europe	75.5%	7	7
University of Copenhagen	Denmark	Europe	72.8%	8	8
KU Leuven	Belgium	Europe	72.1%	9	9
Ecole des Ponts ParisTech	France	Europe	71.2%	10	10
Uppsala University	Sweden	Europe	69.3%	11	11
Ecole Polytechnique Federale de Lausanne	Switzerland	Europe	68.3%	12	12
King Abdulaziz University	Saudi Arabia	Asia	67.6%	13	1
Ecole Polytechnique	France	Europe	67.3%	14	13
ETH Zurich	Switzerland	Europe	66.1%	15	14
Imperial College London	UK	Europe	66.0%	16	15
Delft University of Technology	Netherlands	Europe	65.5%	17	16
Pohang University of Science & Technology (POSTECH)	Korea	Asia	65.2%	18	2
Lund University	Sweden	Europe	64.9%	19	17
Universite Catholique Louvain	Belgium	Europe	64.4%	20	18
RWTH Aachen University	Germany	Europe	62.5%	21	19

Table 14: (*Continued*)

University	Location	Region	Collaborative-ness score	Collaborat. global rank	Collaborat. regional rank
Humboldt University of Berlin	Germany	Europe	62.4%	22	20
Aalto University	Finland	Europe	62.3%	23	21
Wageningen University & Research	Netherlands	Europe	62.2%	24	22
Eberhard Karls University of Tubingen	Germany	Europe	62.0%	25	23
Technical University of Munich	Germany	Europe	61.7%	26	24
Free University of Berlin	Germany	Europe	61.6%	27	25
University of Antwerp	Belgium	Europe	61.6%	28	26
Nanyang Technological University	Singapore	Asia	61.3%	29	3
University of Lausanne	Switzerland	Europe	61.2%	30	27
Maastricht University	Netherlands	Europe	61.2%	31	28
University of Geneva	Switzerland	Europe	60.9%	32	29
University of Zurich	Switzerland	Europe	60.9%	33	30
University of Oslo	Norway	Europe	60.9%	34	31
Queen Mary University London	UK	Europe	60.5%	35	32
Dresden University of Technology	Germany	Europe	60.5%	36	33
Vienna University of Technology	Austria	Europe	60.2%	37	34
Ruprecht Karls University Heidelberg	Germany	Europe	60.1%	38	35
Massachusetts Institute of Technology (MIT)	USA	N America	60.0%	39	1
Leiden University	Netherlands	Europe	59.6%	40	36
PSL Research University Paris (ComUE)	France	Europe	59.2%	41	37
Aarhus University	Denmark	Europe	59.0%	42	38
Utrecht University	Netherlands	Europe	58.7%	43	39
Carnegie Mellon University	USA	N America	58.5%	44	2

Table 14: (*Continued*)

University	Location	Region	Collaborative-ness score	Collaborat. global rank	Collaborat. regional rank
University of Cambridge	UK	Europe	58.5%	45	40
Karlsruhe Institute of Technology	Germany	Europe	58.4%	46	41
Ghent University	Belgium	Europe	58.3%	47	42
University of Barcelona	Spain	Europe	57.9%	48	43
University of Bergen	Norway	Europe	57.8%	49	44
Erasmus University Rotterdam	Netherlands	Europe	57.7%	50	45
University of Amsterdam	Netherlands	Europe	57.6%	51	46
University of Groningen	Netherlands	Europe	57.5%	52	47
Sorbonne Universite	France	Europe	57.0%	53	48
University of Bern	Switzerland	Europe	56.9%	54	49
University of Glasgow	UK	Europe	56.8%	55	50
University of Manchester	UK	Europe	56.6%	56	51
University of Twente	Netherlands	Europe	56.4%	57	52
University of Southampton	UK	Europe	56.3%	58	53
Vrije Universiteit Amsterdam	Netherlands	Europe	56.0%	59	54
McMaster University	Canada	N America	55.9%	60	3
University of Helsinki	Finland	Europe	55.8%	61	55
Technical University of Berlin	Germany	Europe	55.7%	62	56
University of Munich	Germany	Europe	55.6%	63	57
Radboud University Nijmegen	Netherlands	Europe	54.8%	64	58
Polytechnic University of Milan	Italy	Europe	54.8%	65	59
University of Cape Town	South Africa	Africa	54.6%	66	1
University of California San Diego	USA	N America	54.5%	67	4

Table 14: (*Continued*)

University	Location	Region	Collaborative-ness score	Collaborat. global rank	Collaborat. regional rank
Vrije Universiteit Brussel	Belgium	Europe	54.5%	68	60
University of Toronto	Canada	N America	54.4%	69	5
National University of Singapore	Singapore	Asia	54.4%	70	4
Georgia Institute of Technology	USA	N America	54.3%	71	6
University of Liverpool	UK	Europe	54.2%	72	61
University of Bonn	Germany	Europe	54.2%	73	62
Princeton University	USA	N America	54.0%	74	7
University of Oxford	UK	Europe	53.9%	75	63
University of Vienna	Austria	Europe	53.8%	76	64
Stockholm University	Sweden	Europe	53.7%	77	65
Stanford University	USA	N America	53.7%	78	8
University of Hamburg	Germany	Europe	53.6%	79	66
University College London	UK	Europe	53.5%	80	67
Scuola Normale Superiore di Pisa	Italy	Europe	53.4%	81	68
American University of Beirut	Lebanon	Asia	53.1%	82	5
University of Edinburgh	UK	Europe	52.9%	83	69
Harvard University	USA	N America	52.8%	84	9
University of Aberdeen	UK	Europe	52.4%	85	70
Korea Advanced Institute of Science & Technology (KAIST)	Korea	Asia	52.4%	86	6
Sungkyunkwan University (SKKU)	Korea	Asia	52.3%	87	7
Ecole Normale Superieure de Lyon (ENS LYON)	France	Europe	52.1%	88	71
Duke University	USA	N America	51.9%	89	10

Table 14: (*Continued*)

University	Location	Region	Collaborative-ness score	Collaborat. global rank	Collaborat. regional rank
University of Sheffield	UK	Europe	51.9%	90	72
McGill University	Canada	N America	51.4%	91	11
University of Leicester	UK	Europe	51.3%	92	73
University of Freiburg	Germany	Europe	51.2%	93	74
University of Leeds	UK	Europe	51.0%	94	75
University of Montreal	Canada	N America	50.8%	95	12
Scuola Superiore Sant'Anna	Italy	Europe	50.8%	96	76
California Institute of Technology	USA	N America	50.8%	97	13
Queens University Belfast	UK	Europe	50.6%	98	77
Kings College London	UK	Europe	50.5%	99	78
University of Canterbury	New Zealand	Oceania	50.2%	100	1
University of California Santa Barbara	USA	N America	50.1%	101	14
University of Bristol	UK	Europe	49.8%	102	79
University of British Columbia	Canada	N America	49.6%	103	15
University of Waterloo	Canada	N America	49.5%	104	16
University of the Andes Colombia	Colombia	S America	48.7%	105	1
Autonomous University of Barcelona	Spain	Europe	48.6%	106	80
University of Alberta	Canada	N America	48.6%	107	17
University College Dublin	Ireland	Europe	48.5%	108	81
Newcastle University - UK	UK	Europe	48.5%	109	82
University of Bologna	Italy	Europe	48.5%	110	83
Tokyo Institute of Technology	Japan	Asia	48.5%	111	8
University of York - UK	UK	Europe	48.4%	112	84

Table 14: (*Continued*)

University	Location	Region	Collaborative-ness score	Collaborat. global rank	Collaborat. regional rank
Waseda University	Japan	Asia	48.3%	113	9
University of Bath	UK	Europe	48.3%	114	85
University of California Los Angeles	USA	N America	48.3%	115	18
University of Western Australia	Australia	Oceania	48.2%	116	2
University of Auckland	New Zealand	Oceania	48.1%	117	3
University of Gottingen	Germany	Europe	47.8%	118	86
Cardiff University	UK	Europe	47.4%	119	87
University of Navarra	Spain	Europe	47.3%	120	88
University of California Berkeley	USA	N America	47.3%	121	19
Pontificia Universidad Catolica de Chile	Chile	S America	47.2%	122	2
University of Calgary	Canada	N America	47.0%	123	20
Johns Hopkins University	USA	N America	47.0%	124	21
University of Washington Seattle	USA	N America	46.9%	125	22
University of Technology Sydney	Australia	Oceania	46.9%	126	4
University of Birmingham	UK	Europe	46.8%	127	89
University of Padua	Italy	Europe	46.6%	128	90
Trinity College Dublin	Ireland	Europe	46.6%	129	91
Lancaster University	UK	Europe	46.2%	130	92
University of Sydney	Australia	Oceania	46.0%	131	5
University of Otago	New Zealand	Oceania	46.0%	132	6
University of Southern California	USA	N America	45.4%	133	23
Columbia University	USA	N America	45.4%	134	24
Tohoku University	Japan	Asia	45.3%	135	10

Table 14: (*Continued*)

University	Location	Region	Collaborative-ness score	Collaborat. global rank	Collaborat. regional rank
Tel Aviv University	Israel	Asia	45.0%	136	11
University of Nottingham	UK	Europe	44.8%	137	93
University of Rochester	USA	N America	44.8%	138	25
Hebrew University of Jerusalem	Israel	Asia	44.8%	139	12
University of Melbourne	Australia	Oceania	44.7%	140	7
Universidad de Chile	Chile	S America	44.6%	141	3
University of Texas Austin	USA	N America	44.4%	142	26
Rice University	USA	N America	44.3%	143	27
Curtin University	Australia	Oceania	44.3%	144	8
Autonomous University of Madrid	Spain	Europe	44.3%	145	94
University of Reading	UK	Europe	44.3%	146	95
University of Warwick	UK	Europe	44.2%	147	96
University of Queensland	Australia	Oceania	43.8%	148	9
Boston University	USA	N America	43.8%	149	28
University of Maryland College Park	USA	N America	43.8%	150	29
University of Adelaide	Australia	Oceania	43.7%	151	10
University of St Andrews	UK	Europe	43.7%	152	97
Purdue University	USA	N America	43.6%	153	30
Seoul National University (SNU)	Korea	Asia	43.6%	154	13
University of New South Wales Sydney	Australia	Oceania	43.5%	155	11
Cornell University	USA	N America	43.5%	156	31
University of Exeter	UK	Europe	43.2%	157	98
University of Tokyo	Japan	Asia	43.2%	158	14

Table 14: (*Continued*)

University	Location	Region	Collaborative-ness score	Collaborat. global rank	Collaborat. regional rank
Monash University	Australia	Oceania	43.1%	159	12
Chinese University of Hong Kong	Hong Kong	Asia	43.1%	160	15
Loughborough University	UK	Europe	43.1%	161	99
University of Chicago	USA	N America	43.0%	162	32
University of Wollongong	Australia	Oceania	42.5%	163	13
Durham University	UK	Europe	42.3%	164	100
Osaka University	Japan	Asia	42.2%	165	16
Australian National University	Australia	Oceania	42.1%	166	14
Chulalongkorn University	Thailand	Asia	42.0%	167	17
University of Sussex	UK	Europe	41.9%	168	101
King Fahd University of Petroleum & Minerals	Saudi Arabia	Asia	41.8%	169	18
University of California Irvine	USA	N America	41.8%	170	33
Western University (University of Western Ontario)	Canada	N America	41.8%	171	34
University of Illinois Urbana-Champaign	USA	N America	41.7%	172	35
University of Pennsylvania	USA	N America	41.6%	173	36
University of Buenos Aires	Argentina	S America	41.1%	174	4
University of Colorado Boulder	USA	N America	41.0%	175	37
University of California Davis	USA	N America	41.0%	176	38
Northwestern University	USA	N America	40.9%	177	39
Hong Kong University of Science & Technology	Hong Kong	Asia	40.5%	178	19
Emory University	USA	N America	40.5%	179	40
University of North Carolina Chapel Hill	USA	N America	40.4%	180	41
Tecnologico de Monterrey	Mexico	N America	40.4%	181	42

Table 14: (*Continued*)

University	Location	Region	Collaborative-ness score	Collaborat. global rank	Collaborat. regional rank
Texas A&M University College Station	USA	N America	40.4%	182	43
Sapienza University Rome	Italy	Europe	40.4%	183	102
Yale University	USA	N America	40.3%	184	44
Macquarie University	Australia	Oceania	40.3%	185	15
University of Wisconsin Madison	USA	N America	40.3%	186	45
Victoria University Wellington	New Zealand	Oceania	40.3%	187	16
Vanderbilt University	USA	N America	40.2%	188	46
Queensland University of Technology (QUT)	Australia	Oceania	40.0%	189	17
Hanyang University	Korea	Asia	40.0%	190	20
Queens University - Canada	Canada	N America	39.8%	191	47
University of Michigan	USA	N America	39.7%	192	48
Kyushu University	Japan	Asia	39.6%	193	21
Royal Melbourne Institute of Technology (RMIT)	Australia	Oceania	39.6%	194	18
New York University	USA	N America	39.6%	195	49
University of Minnesota Twin Cities	USA	N America	39.5%	196	50
Kyoto University	Japan	Asia	39.4%	197	22
Korea University	Korea	Asia	39.3%	198	23
University of Newcastle	Australia	Oceania	39.0%	199	19
Universiti Malaya	Malaysia	Asia	38.8%	200	24
National Taiwan University	Taiwan	Asia	38.6%	201	25
National Chiao Tung University	Taiwan	Asia	38.5%	202	26
Georgetown University	USA	N America	38.5%	203	51
Yonsei University	Korea	Asia	38.4%	204	27

Table 14: (*Continued*)

University	Location	Region	Collaborative-ness score	Collaborat. global rank	Collaborat. regional rank
Ohio State University	USA	N America	38.2%	205	52
University of Florida	USA	N America	38.2%	206	53
Penn State University	USA	N America	38.1%	207	54
Nagoya University	Japan	Asia	38.1%	208	28
National Tsing Hua University	Taiwan	Asia	38.0%	209	29
Washington University (WUSTL)	USA	N America	38.0%	210	55
University of Hong Kong	Hong Kong	Asia	37.8%	211	30
Tsinghua University	China (Mainland)	Asia	37.2%	212	31
Complutense University of Madrid	Spain	Europe	37.0%	213	103
Peking University	China (Mainland)	Asia	36.8%	214	32
Keio University	Japan	Asia	36.7%	215	33
Hokkaido University	Japan	Asia	36.5%	216	34
Case Western Reserve University	USA	N America	35.9%	217	56
Arizona State University	USA	N America	35.8%	218	57
Dartmouth College	USA	N America	35.7%	219	58
University of Virginia	USA	N America	35.5%	220	59
University of Pittsburgh	USA	N America	35.4%	221	60
Universidad Nacional Autonoma de Mexico	Mexico	N America	34.8%	222	61
Fudan University	China (Mainland)	Asia	34.4%	223	35
Saint Petersburg State University	Russia	Europe	34.2%	224	104
University of Notre Dame	USA	N America	33.9%	225	62
Michigan State University	USA	N America	33.8%	226	63
Shanghai Jiao Tong University	China (Mainland)	Asia	33.5%	227	36

Table 14: (*Continued*)

University	Location	Region	Collaborative-ness score	Collaborat. global rank	Collaborat. regional rank
Brown University	USA	N America	33.2%	228	64
University of Science & Technology of China	China (Mainland)	Asia	32.8%	229	37
City University of Hong Kong	Hong Kong	Asia	32.7%	230	38
Al-Farabi Kazakh National University	Kazakhstan	Asia	32.6%	231	39
Universiti Teknologi Malaysia	Malaysia	Asia	31.8%	232	40
Universiti Sains Malaysia	Malaysia	Asia	31.6%	233	41
Lomonosov Moscow State University	Russia	Europe	31.5%	234	105
Universidade de Sao Paulo	Brazil	S America	30.7%	235	5
Universiti Putra Malaysia	Malaysia	Asia	30.6%	236	42
Indian Institute of Science (IISC) - Bangalore	India	Asia	30.5%	237	43
Novosibirsk State University	Russia	Europe	30.5%	238	106
National Cheng Kung University	Taiwan	Asia	30.5%	239	44
London School Economics & Political Science	UK	Europe	30.2%	240	107
Indian Institute of Technology (IIT) - Bombay	India	Asia	29.9%	241	45
University of Illinois Chicago	USA	N America	29.6%	242	65
Kyung Hee University	Korea	Asia	29.5%	243	46
Hong Kong Polytechnic University	Hong Kong	Asia	29.0%	244	47
Universidade Estadual de Campinas	Brazil	S America	28.9%	245	6
Zhejiang University	China (Mainland)	Asia	28.4%	246	48
Nanjing University	China (Mainland)	Asia	27.9%	247	49
Institut d'Etudes Politiques Paris (SciencePo)	France	Europe	25.0%	248	108
Indian Institute of Technology (IIT) - Delhi	India	Asia	24.4%	249	50
Universiti Kebangsaan Malaysia	Malaysia	Asia	23.5%	250	51

Table 15: Global rankings overview (overall and 3 aspects)

University	Location	Region	Overall rank	Multidiscipl. rank	Impact rank	Collaborat. rank
Massachusetts Institute of Technology (MIT)	USA	N America	1	96	1	39
Technical University of Denmark	Denmark	Europe	2	3	88	5
Ecole Polytechnique Federale de Lausanne	Switzerland	Europe	3	35	18	12
Carnegie Mellon University	USA	N America	4	50	7	44
Eindhoven University of Technology	Netherlands	Europe	5	15	157	3
CentraleSupelec	France	Europe	6	22	205	1
Telecom ParisTech	France	Europe	7	31	172	2
Wageningen University & Research	Netherlands	Europe	8	40	34	24
ETH Zurich	Switzerland	Europe	9	55	26	15
Nanyang Technological University	Singapore	Asia	10	12	60	29
Stanford University	USA	N America	11	190	2	78
Chalmers University of Technology	Sweden	Europe	12	25	185	4
Georgia Institute of Technology	USA	N America	13	17	54	71
Royal Institute of Technology	Sweden	Europe	14	32	151	7
University of California Berkeley	USA	N America	15	108	5	121
Harvard University	USA	N America	16	246	3	84
Princeton University	USA	N America	17	211	6	74
California Institute of Technology	USA	N America	18	241	4	97
KU Leuven	Belgium	Europe	19	89	55	9
Delft University of Technology	Netherlands	Europe	20	5	167	17
Ecole des Ponts ParisTech	France	Europe	21	21	159	10

Table 15: (*Continued*)

University	Location	Region	Overall rank	Multidiscipl. rank	Impact rank	Collaborat. rank
University of California Santa Barbara	USA	N America	22	93	11	101
Aalto University	Finland	Europe	23	29	113	23
King Abdulaziz University	Saudi Arabia	Asia	24	74	71	13
Imperial College London	UK	Europe	25	152	37	16
University of California San Diego	USA	N America	26	184	14	67
Hong Kong University of Science & Technology	Hong Kong	Asia	27	37	35	178
University of Basel	Switzerland	Europe	28	187	79	6
Rice University	USA	N America	29	68	20	143
Pohang University of Science & Technology (POSTECH)	Korea	Asia	30	34	163	18
University of Twente	Netherlands	Europe	31	9	158	57
University of Copenhagen	Denmark	Europe	32	199	64	8
University of Cambridge	UK	Europe	33	182	29	45
Duke University	USA	N America	34	224	13	89
Queen Mary University London	UK	Europe	35	231	27	35
University of Washington Seattle	USA	N America	36	201	12	125
Erasmus University Rotterdam	Netherlands	Europe	37	207	30	50
Polytechnic University of Milan	Italy	Europe	38	1	203	65
University of Chicago	USA	N America	39	228	8	162
National University of Singapore	Singapore	Asia	40	51	97	70
Technical University of Munich	Germany	Europe	41	76	100	26
Scuola Normale Superiore di Pisa	Italy	Europe	42	243	17	81

Table 15: (*Continued*)

University	Location	Region	Overall rank	Multidiscipl. rank	Impact rank	Collaborat. rank
University of Oxford	UK	Europe	43	219	25	75
University of Maryland College Park	USA	N America	44	61	41	150
Utrecht University	Netherlands	Europe	45	118	59	43
Universite Catholique Louvain	Belgium	Europe	46	166	73	20
Uppsala University	Sweden	Europe	47	176	99	11
Radboud University Nijmegen	Netherlands	Europe	48	125	51	64
Vrije Universiteit Amsterdam	Netherlands	Europe	49	106	67	59
Columbia University	USA	N America	50	200	19	134
Scuola Superiore Sant'Anna	Italy	Europe	51	11	178	96
University of Amsterdam	Netherlands	Europe	52	197	45	51
University of Zurich	Switzerland	Europe	53	205	56	33
Vienna University of Technology	Austria	Europe	54	30	192	37
University of California Los Angeles	USA	N America	55	210	24	115
Karlsruhe Institute of Technology	Germany	Europe	56	45	160	46
Korea Advanced Institute of Science & Technology (KAIST)	Korea	Asia	57	14	183	86
University of Edinburgh	UK	Europe	58	220	33	83
Northwestern University	USA	N America	59	186	16	177
University College London	UK	Europe	60	156	44	80
University of Texas Austin	USA	N America	61	70	50	142
PSL Research University Paris (ComUE)	France	Europe	62	179	68	41
Yale University	USA	N America	63	221	10	184

Table 15: (*Continued*)

University	Location	Region	Overall rank	Multidiscipl. rank	Impact rank	Collaborat. rank
University of Technology Sydney	Australia	Oceania	64	26	146	126
Johns Hopkins University	USA	N America	65	230	21	124
Ecole Polytechnique	France	Europe	66	138	119	14
University of Groningen	Netherlands	Europe	67	163	66	52
Stockholm University	Sweden	Europe	68	160	48	77
University of Antwerp	Belgium	Europe	69	79	131	28
Washington University (WUSTL)	USA	N America	70	242	9	210
University of Colorado Boulder	USA	N America	71	101	32	175
University of Pennsylvania	USA	N America	72	212	15	173
City University of Hong Kong	Hong Kong	Asia	73	6	117	230
Technical University of Berlin	Germany	Europe	74	28	190	62
University of Southampton	UK	Europe	75	99	91	58
Lund University	Sweden	Europe	76	148	110	19
Cornell University	USA	N America	77	206	22	156
Leiden University	Netherlands	Europe	78	235	57	40
RWTH Aachen University	Germany	Europe	79	64	171	21
University of Southern California	USA	N America	80	122	43	133
Boston University	USA	N America	81	194	31	149
University of Geneva	Switzerland	Europe	82	239	65	32
Dresden University of Technology	Germany	Europe	83	87	142	36
University of Toronto	Canada	N America	84	227	46	69
Aarhus University	Denmark	Europe	85	127	105	42

Table 15: (*Continued*)

University	Location	Region	Overall rank	Multidiscipl. rank	Impact rank	Collaborat. rank
Vanderbilt University	USA	N America	86	203	23	188
University of North Carolina Chapel Hill	USA	N America	87	168	28	180
Kings College London	UK	Europe	88	153	58	99
University of Illinois Urbana-Champaign	USA	N America	89	59	74	172
University of Barcelona	Spain	Europe	90	188	84	48
Hong Kong Polytechnic University	Hong Kong	Asia	91	4	150	244
Ghent University	Belgium	Europe	92	77	152	47
McMaster University	Canada	N America	93	161	83	60
McGill University	Canada	N America	94	104	90	91
University of Bristol	UK	Europe	95	164	61	102
Humboldt University of Berlin	Germany	Europe	96	217	102	22
University of Waterloo	Canada	N America	97	38	180	104
University of Oslo	Norway	Europe	98	198	104	34
University of Lausanne	Switzerland	Europe	99	229	92	30
Ruprecht Karls University Heidelberg	Germany	Europe	100	249	52	38
University of Sussex	UK	Europe	101	141	42	168
Lancaster University	UK	Europe	102	81	89	130
University of California Irvine	USA	N America	103	172	38	170
University of Glasgow	UK	Europe	104	236	70	55
Maastricht University	Netherlands	Europe	105	159	130	31
Vrije Universiteit Brussel	Belgium	Europe	106	78	149	68
Chinese University of Hong Kong	Hong Kong	Asia	107	113	62	160

Table 15: (*Continued*)

University	Location	Region	Overall rank	Multidiscipl. rank	Impact rank	Collaborat. rank
Tsinghua University	China (Mainland)	Asia	108	24	153	212
University of Manchester	UK	Europe	109	178	103	56
Sorbonne Universite	France	Europe	110	218	86	53
Arizona State University	USA	N America	111	42	109	218
University of Michigan	USA	N America	112	171	40	192
University of British Columbia	Canada	N America	113	121	95	103
University of Helsinki	Finland	Europe	114	173	107	61
University of Munich	Germany	Europe	115	234	77	63
Newcastle University - UK	UK	Europe	116	116	101	109
University of Wollongong	Australia	Oceania	117	33	179	163
Free University of Berlin	Germany	Europe	118	208	132	27
University of Bath	UK	Europe	119	46	177	114
Eberhard Karls University of Tubingen	Germany	Europe	120	240	111	25
University of Bergen	Norway	Europe	121	196	126	49
Emory University	USA	N America	122	233	36	179
New York University	USA	N America	123	216	39	195
Cardiff University	UK	Europe	124	147	94	119
University of Liverpool	UK	Europe	125	223	98	72
University of Aberdeen	UK	Europe	126	149	114	85
University of Bologna	Italy	Europe	127	102	120	110
University of Cape Town	South Africa	Africa	128	136	139	66

Table 15: (Continued)

University	Location	Region	Overall rank	Multidiscipl. rank	Impact rank	Collaborat. rank
Purdue University	USA	N America	129	49	168	153
University of Navarra	Spain	Europe	130	97	124	120
University of Sheffield	UK	Europe	131	112	135	90
University of Wisconsin Madison	USA	N America	132	139	72	186
University of Bern	Switzerland	Europe	133	248	93	54
University of Bonn	Germany	Europe	134	245	82	73
Royal Melbourne Institute of Technology (RMIT)	Australia	Oceania	135	7	208	194
National Tsing Hua University	Taiwan	Asia	136	19	193	209
University of Rochester	USA	N America	137	244	47	138
University of Exeter	UK	Europe	138	150	81	157
University of Pittsburgh	USA	N America	139	157	49	221
University of Birmingham	UK	Europe	140	215	78	127
University of Adelaide	Australia	Oceania	141	109	108	151
University of Minnesota Twin Cities	USA	N America	142	123	80	196
University of Leeds	UK	Europe	143	177	122	94
Autonomous University of Barcelona	Spain	Europe	144	95	154	106
Ohio State University	USA	N America	145	162	69	205
University of New South Wales Sydney	Australia	Oceania	146	60	161	155
University of Nottingham	UK	Europe	147	85	143	137
University of York - UK	UK	Europe	148	115	137	112
University of California Davis	USA	N America	149	142	85	176
University of Montreal	Canada	N America	150	140	145	95

Table 15: (*Continued*)

University	Location	Region	Overall rank	Multidiscipl. rank	Impact rank	Collaborat. rank
University of Freiburg	Germany	Europe	151	225	106	93
University of Queensland	Australia	Oceania	152	88	138	148
Ecole Normale Superieure de Lyon (ENS LYON)	France	Europe	153	180	136	88
Loughborough University	UK	Europe	154	8	215	161
University of Science & Technology of China	China (Mainland)	Asia	155	54	127	229
University of Padua	Italy	Europe	156	146	125	128
University of Melbourne	Australia	Oceania	157	134	115	140
University of Reading	UK	Europe	158	91	147	146
University of Auckland	New Zealand	Oceania	159	80	175	117
Queens University Belfast	UK	Europe	160	117	164	98
University of Leicester	UK	Europe	161	247	96	92
University of Warwick	UK	Europe	162	145	112	147
Brown University	USA	N America	163	151	63	228
Sungkyunkwan University (SKKU)	Korea	Asia	164	92	189	87
Penn State University	USA	N America	165	83	128	207
University of Hamburg	Germany	Europe	166	238	123	79
University of Vienna	Austria	Europe	167	129	176	76
University of Alberta	Canada	N America	168	131	156	107
University of Calgary	Canada	N America	169	143	144	123
University of Sydney	Australia	Oceania	170	165	129	131
University of Gottingen	Germany	Europe	171	174	133	118

Table 15: (*Continued*)

University	Location	Region	Overall rank	Multidiscipl. rank	Impact rank	Collaborat. rank
Australian National University	Australia	Oceania	172	132	118	166
Dartmouth College	USA	N America	173	232	53	219
University of Hong Kong	Hong Kong	Asia	174	82	134	211
University of Canterbury	New Zealand	Oceania	175	73	199	100
Curtin University	Australia	Oceania	176	53	197	144
University of the Andes Colombia	Colombia	S America	177	94	182	105
University of Western Australia	Australia	Oceania	178	158	155	116
Georgetown University	USA	N America	179	202	87	203
London School Economics & Political Science	UK	Europe	180	44	169	240
Trinity College Dublin	Ireland	Europe	181	120	170	129
Case Western Reserve University	USA	N America	182	209	75	217
Monash University	Australia	Oceania	183	98	166	159
University of Virginia	USA	N America	184	213	76	220
University College Dublin	Ireland	Europe	185	195	162	108
Michigan State University	USA	N America	186	86	141	226
Autonomous University of Madrid	Spain	Europe	187	192	148	145
Texas A&M University College Station	USA	N America	188	66	188	182
Durham University	UK	Europe	189	222	116	164
Queensland University of Technology (QUT)	Australia	Oceania	190	56	200	189
Hebrew University of Jerusalem	Israel	Asia	191	133	173	139
University of St Andrews	UK	Europe	192	214	140	152
University of Notre Dame	USA	N America	193	128	121	225

Table 15: (*Continued*)

University	Location	Region	Overall rank	Multidiscipl. rank	Impact rank	Collaborat. rank
National Chiao Tung University	Taiwan	Asia	194	13	223	202
University of Florida	USA	N America	195	119	165	206
National Taiwan University	Taiwan	Asia	196	65	196	201
Tel Aviv University	Israel	Asia	197	154	181	136
Tokyo Institute of Technology	Japan	Asia	198	72	213	111
Seoul National University (SNU)	Korea	Asia	199	100	198	154
Queens University - Canada	Canada	N America	200	105	184	191
Peking University	China (Mainland)	Asia	201	111	174	214
Macquarie University	Australia	Oceania	202	110	191	185
King Fahd University of Petroleum & Minerals	Saudi Arabia	Asia	203	27	234	169
Sapienza University Rome	Italy	Europe	204	126	187	183
Waseda University	Japan	Asia	205	62	221	113
American University of Beirut	Lebanon	Asia	206	155	210	82
University of Otago	New Zealand	Oceania	207	170	201	132
Western University (University of Western Ontario)	Canada	N America	208	130	195	171
Tecnologico de Monterrey	Mexico	N America	209	20	236	181
Hanyang University	Korea	Asia	210	41	224	190
Universiti Teknologi Malaysia	Malaysia	Asia	211	2	243	232
University of Newcastle	Australia	Oceania	212	90	207	199
University of Tokyo	Japan	Asia	213	181	202	158
Victoria University Wellington	New Zealand	Oceania	214	69	214	187

Table 15: (*Continued*)

University	Location	Region	Overall rank	Multidiscipl. rank	Impact rank	Collaborat. rank
Shanghai Jiao Tong University	China (Mainland)	Asia	215	52	211	227
Korea University	Korea	Asia	216	57	217	198
National Cheng Kung University	Taiwan	Asia	217	23	225	239
Indian Institute of Technology (IIT) - Bombay	India	Asia	218	36	227	241
Pontificia Universidad Catolica de Chile	Chile	S America	219	237	206	122
University of Illinois Chicago	USA	N America	220	135	186	242
Zhejiang University	China (Mainland)	Asia	221	48	216	246
Fudan University	China (Mainland)	Asia	222	137	204	223
Tohoku University	Japan	Asia	223	84	228	135
Nanjing University	China (Mainland)	Asia	224	103	194	247
Yonsei University	Korea	Asia	225	124	212	204
Kyoto University	Japan	Asia	226	185	209	197
Universiti Malaya	Malaysia	Asia	227	63	230	200
Indian Institute of Science (IISC) - Bangalore	India	Asia	228	43	226	237
Chulalongkorn University	Thailand	Asia	229	58	237	167
Osaka University	Japan	Asia	230	169	219	165
Complutense University of Madrid	Spain	Europe	231	107	220	213
Institut d'Etudes Politiques Paris (SciencePo)	France	Europe	232	39	233	248
Indian Institute of Technology (IIT) - Delhi	India	Asia	233	10	242	249

Table 15: (*Continued*)

University	Location	Region	Overall rank	Multidiscipl. rank	Impact rank	Collaborat. rank
Kyushu University	Japan	Asia	234	114	229	193
Universiti Sains Malaysia	Malaysia	Asia	235	18	246	233
Universidad de Chile	Chile	S America	236	193	231	141
Nagoya University	Japan	Asia	237	204	218	208
University of Buenos Aires	Argentina	S America	238	175	232	174
Universiti Putra Malaysia	Malaysia	Asia	239	16	249	236
Keio University	Japan	Asia	240	191	222	215
Kyung Hee University	Korea	Asia	241	71	240	243
Hokkaido University	Japan	Asia	242	167	235	216
Universidade Estadual de Campinas	Brazil	S America	243	75	239	245
Universidade de Sao Paulo	Brazil	S America	244	144	241	235
Universiti Kebangsaan Malaysia	Malaysia	Asia	245	47	248	250
Novosibirsk State University	Russia	Europe	246	226	238	238
Universidad Nacional Autonoma de Mexico	Mexico	N America	247	183	244	222
Saint Petersburg State University	Russia	Europe	248	189	247	224
Lomonosov Moscow State University	Russia	Europe	249	250	245	234
Al-Farabi Kazakh National University	Kazakhstan	Asia	250	67	250	231

www.ingramcontent.com/pod-product-compliance
Ingram Content Group UK Ltd.
Pitfield, Milton Keynes, MK11 3LW, UK
UKHW050009160726
7214IPUK00019B/348